# WHEN WILL I EVER TEACH THIS?

## AN ACTIVITIES MANUAL FOR MATHEMATICS FOR ELEMENTARY TEACHERS

## Susie M. Lanier

*Georgia Southern University*

## Sharon E. Taylor

*Georgia Southern University*

Boston   San Francisco   New York
London   Toronto   Sydney   Tokyo   Singapore   Madrid
Mexico City   Munich   Paris   Cape Town   Hong Kong   Montreal

| | |
|---|---|
| Publisher | Greg Tobin |
| Acquisitions Editor | Anne Kelly |
| Editorial Assistant | Cecilia Fleming |
| Senior Author Support/Technology Specialist | Joe Vetere |
| Production Supervisor | Sheila Spinney |
| Cover Designer | Beth Anderson |
| Marketing Manager | Becky Anderson |
| First Print Buyer | Evelyn Beaton |

## Credits

Pps 2, 3 From SCOTT-FORESMAN -ADDISON WESLEY MATH (Grade 1). Copyright © 2002 by Pearson Education, Inc. Reprinted by permission.

Pps 7, 8, 12, 13, 33, 34, 99, 100, 132, 133, 134, 135, 141, 142 From SCOTT-FORESMAN - ADDISON WES-LEY MATH (Grade 2). Copyright © 2002 by Pearson Education, Inc. Reprinted by permission.

Pps 17, 18, 60, 61, 105, 106 From SCOTT-FORESMAN - ADDISON WESLEY MATH (Grade 3). Copyright © 2002 by Pearson Education, Inc. Reprinted by permission.

Pps 22, 23, 50, 51, 65, 66, 107, 108, 117, 118, 122, 123, 127, 128, 136, 137, 143, 144, 148, 149 From SCOTT-FORESMAN - ADDISON WESLEY MATH (Grade 4). Copyright © 2002 by Pearson Education, Inc. Reprinted by permission.

Pps 27, 28, 45, 46, 55, 56, 70, 71, 80, 81, 112, 113 From SCOTT-FORESMAN - ADDISON WESLEY MATH (Grade 5). Copyright © 2002 by Pearson Education, Inc. Reprinted by permission.

Pps 38, 39, 40, 75, 76, 85, 86, 87, 88, 92, 93, 94, 95, 150, 151, 152, 153 From SCOTT-FORESMAN - ADDI-SON WESLEY MATH (Grade 6). Copyright © 2002 by Pearson Education, Inc. Reprinted by permission.

Reproduced by Pearson Addison-Wesley from electronic files supplied by the authors.

Copyright © 2004 Pearson Education, Inc.
Publishing as Pearson Addison-Wesley, 75 Arlington Street, Boston MA 02116

ISBN    0-321-23717-X

3 4 5 6 VHG 06 05

PEARSON
Addison
Wesley

# CONTENTS

Sets............................................................................................ 1

Venn Diagrams.......................................................................... 6

Whole Numbers, Operation, and Algorithms............................................. 11

Ancient Numeration Systems............................................................ 16

Estimation and Rounding................................................................ 21

Least Common Multiple and Greatest Common Factor.................................... 26

Fractions.............................................................................. 32

Addition and Subtraction of Fractions................................................. 37

Multiplication and Division of Fractions.............................................. 44

Mixed Numbers and Improper Fractions.................................................. 49

Mixed Number Operations................................................................ 54

Decimals............................................................................... 59

Operations with Decimals............................................................... 64

Percents............................................................................... 69

Fractions/Decimals/Percents........................................................... 74

Ratio and Proportion................................................................... 79

Describing Data........................................................................ 84

Displaying Data and Interpreting Graphs............................................... 91

Probability............................................................................ 98

Lines and Angles....................................................................... 104

Triangles.............................................................................. 111

Quadrilaterals......................................................................... 116

Polygons............................................................................... 121

Polyhedra.............................................................................. 126

Measurement............................................................................ 131

Perimeter and Area..................................................................... 140

Volume and Surface Area................................................................ 147

# PREFACE

As you may guess from the title, this book is an attempt to answer the most popular question we are asked by our students. Those who aspire to teach early grades wonder why they need to know how to teach fractions, decimals, and percents. Middle grade majors often think it unnecessary to be familiar with alternative strategies and algorithms for addition, subtraction, multiplication, and division. After all, their students will know that already.

Our answer takes several different forms. One answer has nothing to do with mathematics. Sometimes, no matter what grade you plan to teach, the reality of your teaching assignment could be very different. Ask any K–8 teacher. So if you plan to teach early grades, yet end up teaching 5th grade, you need to be aware of what is in the 5th grade and how you can approach teaching it.

Our other answers are more mathematical. No matter what grade level you are teaching, there will be students in your classroom who are in need of additional help and others who need challenging extension activities. You will need to be ready to deal with those students as well as the ones at grade level. But most importantly, all teachers of mathematics at all levels are always laying a foundation for students to use throughout their academic lives. You should be aware of where your students came from and where they are going in the future so that you can best prepare them for their mathematical journey.

The best way we have found to demonstrate to our students the need to learn certain topics is to bring pages from a real K-8 textbook. This allows our students to see when and where a topic occurs in the curriculum and to also see how it is presented in a text. We have done that for you in this book.

Each topic in this book has four components: a brief discussion of the topic, pages from a school text, a set of problems that focus on skills, and a set of problems that focus on concepts. The school text pages are from the Scott-Foresman - Addison Wesley math series. We used Grade 1–6 pages.

The discussion page covers many areas: where the topic is taught in the schools, the text pages we are showing, what to expect in the *Skills* and *Concepts* pages, typical problems our students encounter and how we deal with those problems.

We hope you will find this manual helpful not only in your Mathematics for Elementary Teachers (MET) classes, but as you go into your own classroom. Many skill-builders can be used when you teach. The ideas from the *Concepts* pages can be used for foundations, advanced students, or extra practice for students.

The topics are arranged as you find them in most MET books. These are the topics that our students have the most trouble with. We hope our explanations will help you.

# SETS

Almost every MET book has the topic of sets somewhere near the beginning of the book. In many cases, this is the first topic you see in your mathematics teacher education experience. And in many cases, it is a frustrating topic that you may not feel relates to mathematics or your classroom.

You can see from the school text pages that sets are indeed taught in the K-8 classroom. In fact, the pages shown are from the Grade 1 book. This demonstrates the use of sets as a method for working with addition. The Grade 3 text also uses sets when working with fractions. So while the sets you encounter in your classrooms might not be the same type of things you see in your MET course, there is definitely a need to understand sets and their properties.

When working with sets and set operations, one of the biggest obstacles is the language. When definitions for union use the word "or" and definitions for intersection use "and," mathematicians understand those meanings from a mathematical standpoint. However, most non-mathematicians take a much more literal approach to the terminology, which can lead to misunderstanding and misconceptions. It is for this reason that we try to make the definitions more understandable, not by dismissing the mathematics, but by putting union and intersection in terms you relate to.

When working with intersection, think of a highway intersection. What happens there? Two roads meet. There is a place where the two roads share common ground. The same holds true for sets. This does not change the definition of intersection that appears in your text. It simply changes the perspective. So when working with intersection, think of what the sets have in common.

The union of two sets is like a marriage. In fact, many definitions of marriage include the word union. It is about two people joining together. With sets, you are joining together things that are in two (or more) sets. As with a married couple joining two households together, you do not need to keep more than one of any item. The same holds true for sets. When listing things in a union, you do not have to list an element more than once.

Hopefully, these explanations of union and intersection clear up some misconceptions you might have and will help you work with the attached pages. The *Focus on Skills* sheet provides practice with basic set operations. The activities on the *Focus on Concepts* sheet look at order and grouping related to sets.

Name _____

# Show Addition

**Learn** ● ● ● ● ● ● ● ● ● ● ● ● ● ● ● ● ● ● ● ● ● ● ● ● ● ● ●

| part | | part | | in all |
|------|-----|------|-------|--------|
| 2 | and | 5 | is | 7. |
| 2 | plus | 5 | equals | 7. |
| 2 | + | 5 | = | 7 |

$2 + 5 = 7$
is called a
number sentence.
I can write it as
$7 = 2 + 5$, too!

**Check** ● ● ● ● ● ● ● ● ● ● ● ● ● ● ● ● ● ● ● ● ● ●

You can use ⬤ ⬤ and ▭ . Make two parts.
Write the number sentence.

① $3 + 5 = 8$

② ___ + ___ = ___

③ ___ + ___ = ___

**Talk About It** Tell another way to make 9.

Write a number sentence.

**Notes from Home:** Your child learned about the plus sign ($+$) and the equal sign ($=$).
*Home Activity:* Ask your child to tell what each sign means.

**Practice**

You can use 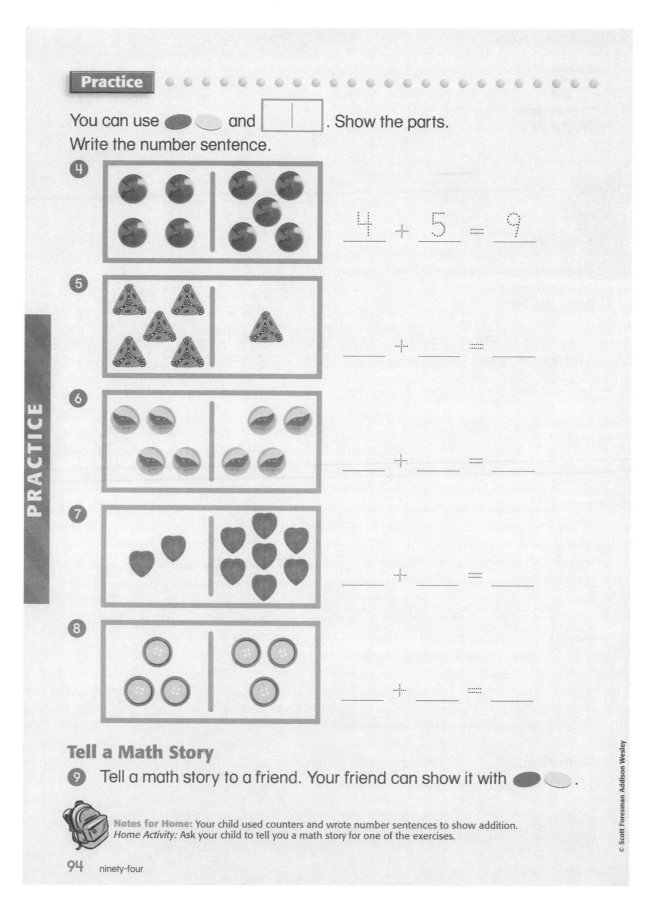 and ☐ . Show the parts.
Write the number sentence.

④

$\underline{4} + \underline{5} = \underline{9}$

⑤

$\underline{\phantom{0}} + \underline{\phantom{0}} = \underline{\phantom{0}}$

⑥

$\underline{\phantom{0}} + \underline{\phantom{0}} = \underline{\phantom{0}}$

⑦

$\underline{\phantom{0}} + \underline{\phantom{0}} = \underline{\phantom{0}}$

⑧

$\underline{\phantom{0}} + \underline{\phantom{0}} = \underline{\phantom{0}}$

**Tell a Math Story**

⑨ Tell a math story to a friend. Your friend can show it with ⬮⬯ .

**Notes for Home:** Your child used counters and wrote number sentences to show addition.
*Home Activity:* Ask your child to tell you a math story for one of the exercises.

4       When Will I Ever Teach This?

**FOCUS ON SKILLS**                          **Name** _____
Sets

Use the following sets to perform the given set operations.

    U = {a, b, c, d, e, f, g, h, i, j, k, r, t, y}
    A = {b, i, r, t, h, d, a, y}
    B = {c, a, k, e}
    C = {c, a, r, d}

1.    $A \cup B =$                           2.    $B \cap C =$

3.    $A' =$                                 4.    $A \cap B =$

5.    $(A \cup B)' =$                        6.    $B - C =$

7.    $C - B =$                              8.    $(A \cap B) \cap C =$

9.    $C' =$                                 10.   $C' \cup A =$

11.   $A - (B \cup C) =$                     12.   $(A - B) \cup C =$

13.   $B' \cap C =$                          14.   $C \cap B'$

**FOCUS ON CONCEPTS**                    **Name** _____
**Sets**

Do set operations have the same properties as other operations?  How can we know this?
The commutative, associative, and identity properties will be explored as they relate to
sets.

1.     Are set operations commutative?  Create two sets A and B and then see what
       happens when you perform the operations.

       a.     Is $A \cap B$ the same as $B \cap A$?


       b.     Is $A \cup B$ the same as $A \cup B$?


       c.     Is $A - B$ the same as $B - A$?


       d.     What might this mean about the operations of addition and subtraction?


2.     Are set operations associative?  Create three sets A, B, and C and then see what
       happens when you perform the operations.

       a.     Are $(A \cap B) \cap C$ and $A \cap (B \cap C)$ the same?


       b.     Are $(A \cup B) \cup C$ and $A \cup (B \cup C)$ the same?


       c.     Are $(A - B) - C$ and $A - (B - C)$ the same?


       d.     What might this mean about the operations of addition and subtraction?


3.     The identity element for addition is 0.  This means that if you add 0 to any
       number, the number does not change.  Is there a set, such that if you perform a
       union or intersection with that set, the original set does not change?  Explain your
       answer.

## VENN DIAGRAMS

Venn diagrams have so many uses other than as a visual representation of set operations. Their use in logic, least common factor and greatest common multiple, and probability are only a few reasons why it is important to understand the fundamental ideas.

Many times the topic of Venn diagrams places students in two distinct categories: you like them because you are a visual learner and seeing the sets makes things understandable OR you do not like them because the circles are confusing and you like things listed better. For students who like things listed better than pictures, we have started using the following as a starting point when working with Venn diagrams.

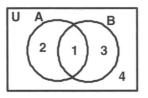

Set A includes regions 1 and 2. Set B includes region 1 and 3. The universal set is regions 1, 2, 3, and 4. So if you are trying to shade the problem $A \cap B$, you want to shade where A and B overlap. By looking at the regions for A (1 and 2) and B (1 and 3), you can see that where they overlap is in region 1. So you should shade region 1.

A similar template for three sets is shown below.

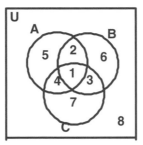

Sets and Venn diagrams first appear in the Grade 1 book and can be found through the Grade 5 book when working with GCF. The pages shown are from the Grade 2 book. In each of these texts, Venn diagrams are not presented simply as a stand-alone topic, but as a way to use the idea to solve other problems. As mentioned earlier, Venn diagrams are a useful tool in many areas.

In *Focus on Skills*, the emphasis is simply on gaining practice with shading Venn diagrams. Although the above templates are not given, you may wish to use them as a starting point. The *Focus on Concepts* page demonstrates the DeMorgan laws as well as how one shaded region can represent two different set operations.

Name _____

# Problem Solving: Collect and Use Data

**Learn** ● ● ● ● ● ● ● ● ● ● ● ● ● ● 

These children showed their
data using a diagram.

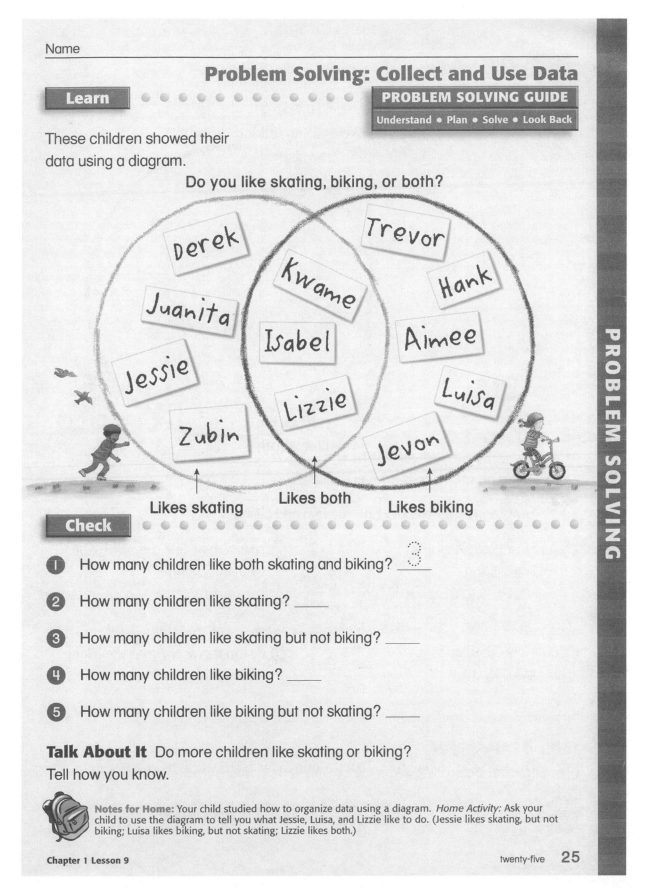

Do you like skating, biking, or both?

Likes skating    Likes both    Likes biking

**Check** ● ● ● ● ● ● ● ● ● ● ● ● ● ● ● ● ● ● ● ● ● ● ● ● ●

1. How many children like both skating and biking? 3

2. How many children like skating? _____

3. How many children like skating but not biking? _____

4. How many children like biking? _____

5. How many children like biking but not skating? _____

**Talk About It** Do more children like skating or biking?
Tell how you know.

**Notes for Home:** Your child studied how to organize data using a diagram. *Home Activity:* Ask your child to use the diagram to tell you what Jessie, Luisa, and Lizzie like to do. (Jessie likes skating, but not biking; Luisa likes biking, but not skating; Lizzie likes both.)

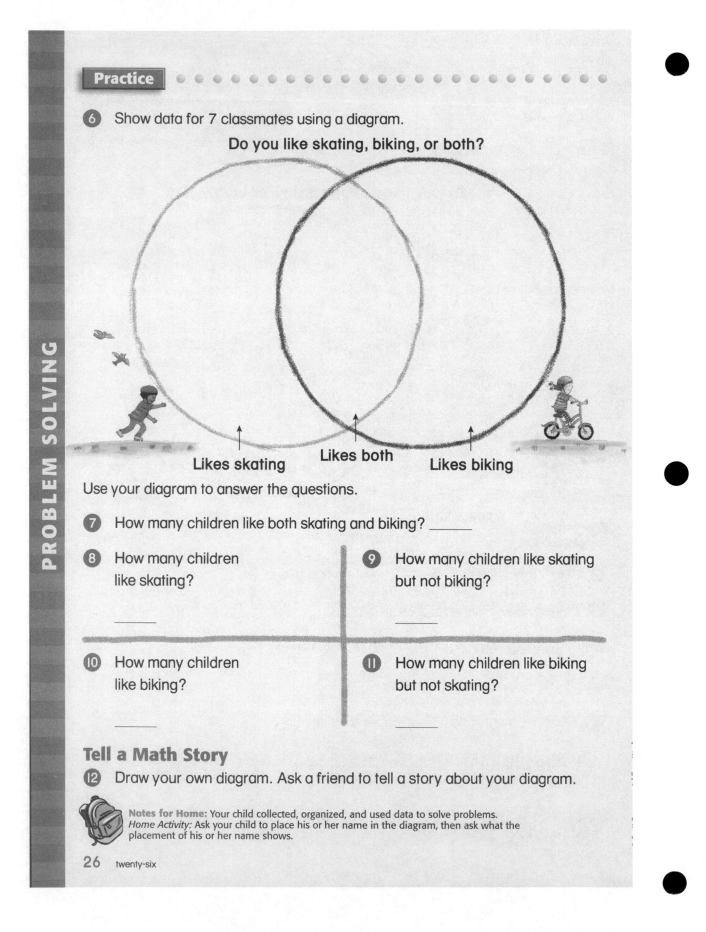

**Practice** ● ● ● ● ● ● ● ● ● ● ● ● ● ● ● ● ● ● ● ● ● ● ● ● ● ● ●

**6** Show data for 7 classmates using a diagram.

**Do you like skating, biking, or both?**

↑ **Likes skating**    ↑ **Likes both**    ↑ **Likes biking**

Use your diagram to answer the questions.

**7** How many children like both skating and biking? _____

**8** How many children like skating?

_____

**9** How many children like skating but not biking?

_____

**10** How many children like biking?

_____

**11** How many children like biking but not skating?

_____

## Tell a Math Story

**12** Draw your own diagram. Ask a friend to tell a story about your diagram.

**Notes for Home:** Your child collected, organized, and used data to solve problems.
*Home Activity:* Ask your child to place his or her name in the diagram, then ask what the placement of his or her name shows.

**FOCUS ON SKILLS**            **Name** _____
**Venn diagrams**

Shade a Venn diagram for each of the following operations.

1.      $A \cup B$                         2.      $A - B$

3.      $B - A$                            4.      $B \cap A$

5.      $A' - B$                           6.      $A' \cup B'$

7.      $(A \cup B) \cap C$               8.      $A \cap (B \cup C)$

Given the following Venn diagrams, write a set operation.

9.  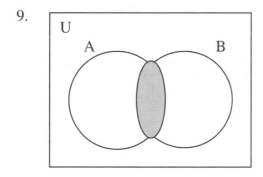                10.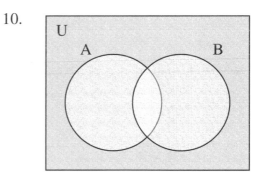

11.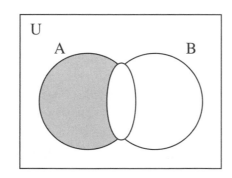

**FOCUS ON CONCEPTS**                    **Name** _____
**Venn Diagrams**


1.    Shade a Venn diagram for $(A \cap B)'$.  On a different diagram, shade $A' \cup B'$.


2.    What happened in these two diagrams?


3.    Shade $(A \cup B)'$.  On a different diagram, shade $A' \cap B'$.


4.    What happened in these two diagrams?


5.    In the previous problems, you have discovered what happens as a result of the DeMorgan laws.  Write a paragraph to describe in everyday language what the DeMorgan law says.


6.    Refer to the Venn diagram in #10 on the *Focus on Skills* sheet.  You should have a set operation to describe the diagram.  Write at least one more set operations to describe this shading.


7.    Write a few sentences to describe in everyday language why it is possible to have two different expressions for the same diagram.

## WHOLE NUMBERS, OPERATIONS, AND ALGORITHMS

If you are anything like our students, you do not want someone telling you how to add, subtract, multiply or divide whole numbers because you learned that a long time ago. You also do not want anyone giving you a lot of alternative methods for working a problem because you know how to do things your way and other ways are confusing. This may not be true for you, but it certainly holds true for a large percentage of our students.

We try to emphasize that there is more than one way to do a problem, whether working with whole numbers or doing any kind of mathematics. By using this approach, we stress the importance of being able to offer alternative strategies to K-8 students who are having trouble. It will not only benefit your students when you begin teaching if you can present different strategies, but will help you with your own problem solving skills along the way.

Our students are quite surprised when we show them that some of these "alternative methods" like regrouping for addition and subtraction, the array model for multiplication, sharing and equal groups for division, and the commutative and associative properties appear in the textbooks they will be teaching from. Regrouping for addition and subtraction begin to appear in the texts at Grade 2 and continues through Grade 5. Multiplication arrays, equal groups, and the sharing model also first appear in the Grade 2 book. Division as repeated subtraction is in the Grade 3 book. Commutative and associative properties are in Grade 2 – Grade 6 books.

The pages shown are from Grade 2 book. As early as second grade, students are exposed to strategies that are not necessarily connected to a formal algorithm for solving a problem. The students you teach will be learning a variety of techniques for looking at the same problem. For this and many other reasons, you must be able to view things from a perspective other than the one you were taught.

One reason many elementary and middle grade students have trouble with mathematics is due in part to only being taught an algorithm with no meaning behind it. This may be the case with you. If you had a negative mathematics experience because you were only taught in a formal and rigid way, then you owe it to your students to seek alternative methods of teaching so that you do not impart negative experiences on your students. These are concepts and skills that elementary students needs for the remainder of their academic careers. You must realize the importance of laying a solid foundation for your students to build on.

The *Focus on Skills* page simply asks you to perform basic problems in a variety of ways. The *Focus on Concepts* page asks you to explain why some algorithms work and how they would explain those to your students. Communication, verbal and written, is an essential skill in teaching all subjects. So yes, there is writing in mathematics.

Name _____

# Explore Addition With or Without Regrouping

**Explore** • • • • • • • • • • • • • • • • • • • • • • • • • •

Do this activity with a partner.

**1** Use cards numbered 1–9. Put the cards facedown.

**2** Take turns. Draw a card and take that many 🟦. Then put the number card back.

**3** Continue taking turns. Trade for a ▭ whenever you can. Keep your ▭ and 🟦 on the chart.

**4** The first player to get 100 wins!

| tens | ones |
|------|------|
|      |      |

**Share** • • • • • • • • • • • • • • • • • • • • • • • • • •

How did you know when to trade 🟦 for ▭ ?

**Notes for Home:** Your child did an activity to group tens and ones. *Home Activity:* Ask your child how to group 7 ones and 5 ones as tens and ones. (1 ten 2 ones)

**Chapter 8 Lesson 6**                    two hundred eighty-one  **281**

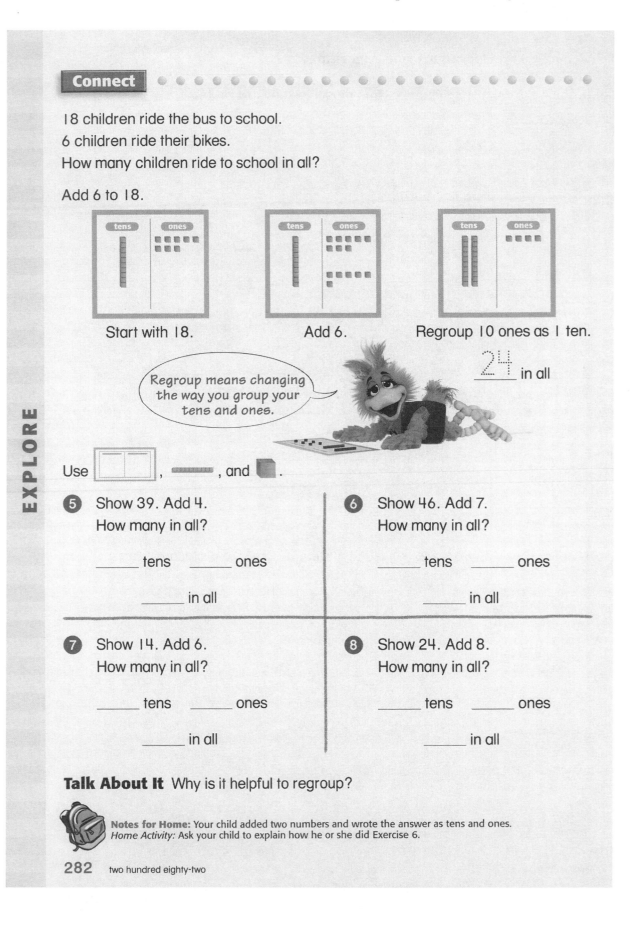

**Connect** • • • • • • • • • • • • • • • • • • • • • • • • • • • •

18 children ride the bus to school.

6 children ride their bikes.

How many children ride to school in all?

Add 6 to 18.

| tens | ones |
| tens | ones |
| tens | ones |

Start with 18.         Add 6.         Regroup 10 ones as 1 ten.

Regroup means changing the way you group your tens and ones.

24 in all

Use [   |   ] , ▭▭▭ , and ▪ .

⑤ Show 39. Add 4.
How many in all?

_____ tens   _____ ones

_____ in all

⑥ Show 46. Add 7.
How many in all?

_____ tens   _____ ones

_____ in all

⑦ Show 14. Add 6.
How many in all?

_____ tens   _____ ones

_____ in all

⑧ Show 24. Add 8.
How many in all?

_____ tens   _____ ones

_____ in all

**Talk About It** Why is it helpful to regroup?

**Notes for Home:** Your child added two numbers and wrote the answer as tens and ones.
*Home Activity:* Ask your child to explain how he or she did Exercise 6.

282    two hundred eighty-two

EXPLORE

**FOCUS ON SKILLS**                          Name _____
**Whole Numbers, Operations, and Algorithms**

For problems 1 – 6, use the number line model to show the problem.

1.      $3 + 4$                              2.      $4 + 2$

3.      $1 + 6$                              4.      $7 - 3$

5.      $8 - 5$                              6.      $3 - 2$

For problems 7 and 8, use the array model to show the problem.

7.      $7 \times 8$                         8.      $13 \times 8$

For problems 9 – 12, use the expanded algorithm to solve the problems.

9.      $427 + 918$                          10.     $3214 + 473$

11.     $381 - 117$                          12.     $512 - 333$

13.     $327 \times 83$                      14.     $419 \times 29$

15.     $5217 \div 37$                       16.     $4758 \div 61$

**FOCUS ON CONCEPTS**                    **Name** _____
**Whole Numbers, Operations, and Algorithms**

1.  Explain why the number line model for addition or subtraction is not the best model for larger numbers.  Discuss another method for addition and subtraction that works better for larger numbers and why it is better.

2.  Give two advantages and two disadvantages for the expanded algorithm for addition.

3.  Give two advantages and two disadvantages for the expanded algorithm for subtraction.

4.  Describe how the base 10 block model for multiplication corresponds to the expanded algorithm (partial products).  Be specific in your explanation.

5.  Give two advantages and two disadvantages for the expanded algorithm for division.

6.  Describe the differences and similarities between the sharing model and the measurement model for division.

## ANCIENT NUMERATION SYSTEMS

One of the most fascinating parts of an MET course is the study of ancient systems. It is a great opportunity to look at history, not just of mathematics, but in general. Looking at systems with respect to place value, the concept of zero, and the symbols that were used is very useful as those ideas relate to our numeration system. Many students do not realize that a thorough understanding of ancient systems can provide an excellent foundation for our system and how to teach it.

Many of our students wonder why it is necessary to bother with ancient systems when these are not used anymore and we only teach our system in the schools. One very obvious reason is because it is in the school texts. However, all the reasons mentioned above are also excellent motivation for studying these systems.

The school text pages are from the Grade 3 book. There are also references to ancient systems in the teacher notes portions of the texts from grades 2 through 6. Not only can the topics be discussed for their mathematical significance, but you can also integrate the ideas into a social studies lesson with a multicultural approach. Looking at the mathematics of an ancient system, such as the Egyptian system, along with the culture and history of Egypt makes for a perfect integrated lesson.

Some books discuss a wide variety of not only ancient systems, but also of modern systems which use symbols different from our own. Again, a focus on a multi-cultural and integrated lesson is an added benefit to a traditional mathematics lesson.

*Focus on Skills* allows for practice with Babylonian, Roman, and Egyptian systems. Some MET books discuss other systems, but these seem to be the most common. We have provided the symbols for you as a reference. We tell our students not to worry too much about precision in re-creating the symbols. However, for students who enjoy artistic creations, drawing the symbols is fun and they really enjoy the art involved.

*Focus on Concepts* is mainly a writing assignment. Comparing and contrasting systems allows for a greater understanding of each system. By asking you to create your own numeration system and describe its features, it is a nice way to see if you understand the concepts. You also get to let your imagination and creativity be the focus of an assignment rather than just your mathematics skills.

# Math Magazine

**Clock Talk** Today we use digits like 1, 2, and 3 to name numbers. Many years ago, pictures of objects or other symbols were used to name numbers. Ancient Romans used the symbols shown below.

| Roman Numeral | I | II | III | IV | V | VI | VII | VIII | IX | X | XI | XII |
|---|---|---|---|---|---|---|---|---|---|---|---|---|
| Standard Form | 1 | 2 | 3 | 4 | 5 | 6 | 7 | 8 | 9 | 10 | 11 | 12 |

The table shows Roman numerals for 1–12.

Here's the key:   $I = 1$

$V = 5$

$X = 10$

Here's what happens when I comes after V or X.

VI  Think of this as
$V + I$   $5 + 1 = 6$

XI  Think of this as
$X + I$   $10 + 1 = 11$

Here's what happens when I comes before V or X.

IV  Think of this as $V - I$   $5 - 1 = 4$

IX  Think of this as $X - I$   $10 - 1 = 9$

You might see Roman numerals on a watch.

▶ **Try These!**
Write each Roman numeral in standard form.
1. XV   2. XIII   3. XVI   4. XX

# Math Magazine

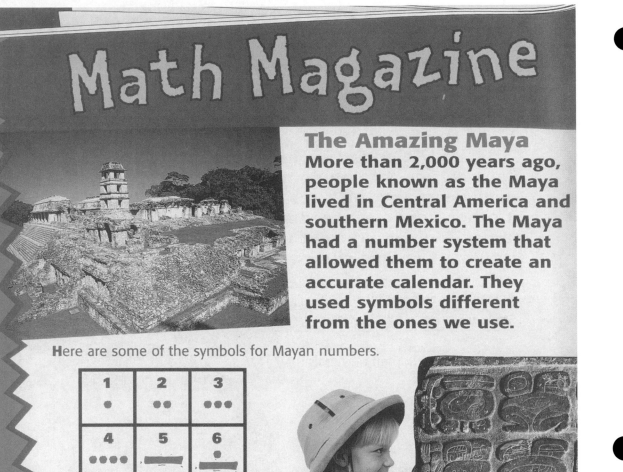

## The Amazing Maya

More than 2,000 years ago, people known as the Maya lived in Central America and southern Mexico. The Maya had a number system that allowed them to create an accurate calendar. They used symbols different from the ones we use.

Here are some of the symbols for Mayan numbers.

| 1 | 2 | 3 |
| 4 | 5 | 6 |
| 7 | 8 | 9 |
| 10 | 11 | 15 |

Can you figure out the key to Mayan numbers? Use the pattern in the table to try.

## Try These!

Write the Mayan symbol for each number.

1. 12    2. 14    3. 18    4. 19

5. **Reasoning** How does skip counting by 5s help you understand Mayan numbers?

**FOCUS ON SKILLS**                                Name _____
**Ancient Numeration Systems**

For problems 1 – 6, change each symbol into a number in our system.

1.     ℙ ℙ 9 9 9 9 9 9 ∩ ∩ ∩ ∩ I I I

2.     ◀ ◀ ▼ ▼ ▼

3.    CLXVI

4.     ☙ ☙ ∩ ∩

5.     ◀ ▼    ▼ ▼ ▼

6.    CDXLVIII

For problems 7 - 9, convert each number to Egyptian, Babylonian, and Roman.

7.    437

8.    9731

9.    889

**FOCUS ON CONCEPTS**                              **Name** _____
**Ancient Numeration Systems**

For problems 1 and 2, compare the two systems listed.  Include things like the use of symbols, place value, the use of zero, and other aspects you consider interesting.

1.    Egyptian and Roman

2.    Egyptian and Babylonian

3.    Given the symbol    ◀ ▼ ▼ ▼    how do you know what number this is?
      Discuss why the Babylonian system can lead to confusion.

4.    Find three places where you are likely to see Roman numerals used today.

5.    Create your own numeration system.  Give all the details of your system.  Then
      write the numbers 497, 3821, and 130 in your system.  Be sure to explain how the
      system works for these conversions and all the steps involved.

# ESTIMATION AND ROUNDING

Estimation and rounding are such important topics throughout all aspects of mathematics, yet many students have trouble. The discomfort level does not seem to be with rounding to a particular place value, although that is true for some, but with the basic skills involved. The lack of estimation skills means students are not able to judge the reasonableness of an answer.

For example, when presented with the problem $\frac{1}{3} + \frac{3}{4}$, do you know if the answer should be more than one or less than one? When this problem is presented by itself, you may not think estimating an answer is necessary. If you want the answer, you can just compute it.

But suppose you are taking a standardized test with four choices given to you. You can save time by knowing what the answer should be close to and eliminating choices that are unreasonable. Now suppose this is the calculation needed to solve a word problem. If you make a calculation error and end up with a strange looking answer, estimation could help you determine if that strange looking answer was close to being correct or not. These strategies are not only important to you, but also for your students.

Estimation is not just a skill to learn in a particular section of the text. Estimation is an invaluable tool to be used to decrease the number of mistakes, a great test-taking strategy for multiple choice tests, and a necessary skill you must be able to teach.

When dealing with whole numbers or decimals (rather than fractions), rounding is a very good way to estimate an answer. For this reason, many times you will see estimation and rounding paired together not only in the texts you teach from, but probably also in the MET book you are using.

Estimation and rounding are first seen in the Grade 2 book and the ideas are revisited in texts through Grade 6. The accompanying pages are from the Grade 4 book. It deals with front-end estimation which is a combination of estimation and rounding. Other books have pages that relate to fractions, decimals, and when it is appropriate to estimate. The fact that these topics appear often is an indication of their importance in the elementary and middle grades classrooms.

*Focus on Skills* does just that. The page is designed to have you practice the basics of rounding and estimation. The *Focus on Concepts* page requires you to explain and justify rounding and estimation strategies.

## Chapter 3
### Lesson 6

# Column Addition

**You Will Learn**
how to add
3 or 4 addends

**Vocabulary**
**addends**
numbers that are
added together
to make a sum

**front-end**
**estimation**
a way to estimate
by first looking at
the leading digits

**Learn** • • • • • • • • • • •

"Don't get in over your head,"
reads Jonathan's ad. His ad
reaches many people who
don't know how to use
computers. Jonathan prints
442 flyers, 339 brochures, and
628 bookmarks. How many
items does he print in all?

Jonathan runs a desktop publishing business
in Farmersville, Illinois.

You can add three **addends** to find the total number of items.
Use **front-end estimation** to check.

**Example**
Find 442 + 339 + 628.

| Step 1 | Step 2 | Step 3 |
|---|---|---|
| Add the ones. Regroup as needed. | Add the tens. Regroup as needed. | Add the hundreds. |

Step 1:
```
  1
 442
 339
+628
   9
```

Step 2:
```
 1 1
 442
 339
+628
  09
```

Step 3:
```
 1 1
  442
  339
+ 628
1,409
```

Round to the front-end,
or leading, digits.

```
442  ⟶   400
339  ⟶   300
+628 ⟶ + 600
       1,300
```

Adjust your estimate.

```
442  ⟶   40
339  ⟶   40
+628 ⟶ + 30
        110
```

1,300 + 110 = 1,410

Since 1,409 is close to 1,410, the answer is reasonable.

Jonathan needs to print 1,409 items.

**Talk About It**

Why do you line up the ones, tens, and hundreds before you add?

## Check

Find each sum. Estimate to check.

| 1. | 2. | 3. | 4. | 5. |
|---|---|---|---|---|
| 32 | 472 | 984 | 75 | 3,004 |
| 41 | 208 | 27 | 348 | 12 |
| + 26 | + 325 | + 3,845 | + 590 | + 8,984 |

6. **Reasoning** When you add three or more addends, does it matter if you change the order of the addends? Explain.

## Practice

### Skills and Reasoning

Find each sum. Estimate to check.

| 7. | 8. | 9. | 10. | 11. |
|---|---|---|---|---|
| 51 | 487 | 280 | 3,298 | 7,855 |
| 41 | 564 | 85 | 2,408 | 4,080 |
| + 85 | + 812 | + 946 | + 6,091 | + 28 |

12. $43 + 27 + 74$    13. $205 + 398 + 190$    14. $147 + 2,490 + 3,580$

15. Find the sum of 2,220 and 540 and 217.

16. Write this number sentence in another way so it has the same sum. $245 + 678 + 2,503 = 3,426$

**Mental Math** Write >, <, or =. Decide without finding the sum.

17. $59 + 34 + 82 \bullet 49 + 24 + 72$    18. $422 + 659 + 394 \bullet 427 + 664 + 402$

### Problem Solving and Applications

19. Suppose Jonathan prints 35 forms, 148 brochures, and 268 flyers. How many items does he print?

20. **Money** Jonathan charged $25 for flyers, $18 for charts, and $22 for banners. What was the total?

21. **Logic** If you use one sheet of paper in an envelope, how many different color combinations can you make?

### Mixed Review and Test Prep

**Mental Math** Find each sum or difference.

22. $6,000 + 5,000$    23. $20,000 - 10,000$    24. $70,000 - 20,000$

25. Find the number that means 3 hundreds, 8 tens, and 6 ones.
Ⓐ 3,086    Ⓑ 317    Ⓒ 386    Ⓓ not here

**FOCUS ON SKILLS**    Name _____
**Estimation and Rounding**

1.    Round 8753 to the nearest

    a.    tens

    b.    hundreds

    c.    thousands

2.    Round 4395 to the nearest

    a.    tens

    b.    hundreds

    c.    thousands

For problems 3- 6, estimate the answer to the nearest hundred.

3.    $371 + 485$    4.    $132 + 505$

5.    $1387 - 725$    6.    $973 - 491$

For problems 7 – 10, estimate the answer to the nearest thousand.

7.    $47 \times 58$    8.    $32 \times 62$

9.    $121 \times 97$    10.    $387 \times 21$

For problems 11 – 14, estimate the answer to the nearest ten.

11.    $983 \div 12$    12.    $1473 \div 51$

13.    $893 \div 33$    14.    $1818 \div 28$

**FOCUS ON CONCEPTS**                    Name _____
**Estimation and Rounding**

Questions 1 – 4 relate to the problems worked on the *Focus on Skills* page. If you have not already worked those problems, go back and do them now so that you can answer the following questions.

1.      In problem 1, you rounded 8753 to the tens, hundreds, and thousands place. Give an example of when each situation would be appropriate. Explain why rounding to that place is appropriate for the situation you are giving.

2.      How would your strategy for problems 3 – 6 change if you were asked to estimate to the nearest ten? Explain your answer.

3.      Explain your estimation strategy for problems 7 – 10. Be specific in describing your thought process.

4.      How would your strategy for problems 11 – 14 change if you were asked to find an exact answer rather than an estimate? Explain your answer.

5.      Why is 300 not an appropriate answer for the problem $4.7 \times 5.8$? What is a good estimate? How did your strategy change for dealing with a decimal problem? How is your strategy the same as before?

## LEAST COMMON MULTIPLE AND GREATEST COMMON FACTOR

For the most part, these topics appear in the school texts as a foundation for working with fractions. Granted, fractions are the main reason these concepts are important. However, LCM and GCF have so many interesting applications on their own. Also, there are many misconceptions about the topics. To illustrate some of the applications and attempt to clear up misconceptions, we spend a good deal of time on these topics.

When teaching the algorithms for finding LCM and GCF, we usually present several alternatives. There is usually a student in the class who uses the same process, but describes it differently. We pursue these avenues in hopes that other students will find it useful. Maybe you have a way of working these problems that is similar to what is taught. If you are not sure, be sure to ask your instructor.

The most common problem occurs because of the name of the concepts. When working with LCM, students tend to focus on the "least" rather than the "multiple" in the title. Remember that multiples of a number are larger than the number itself. So if you are looking for a least common *multiple*, you need to be looking for an answer that is at least as large as the numbers in the problem.

In working with GCF, the same concept applies. Focus on "factor" rather than "greatest". Factors are less than a number. So your answer should be less than or equal to the numbers in your problem. By focusing on the correct word, a lot of the misconceptions go away.

You may be saying that computing LCM and GCF are not a problem, but dealing with those topics in the context of a word problem is hard. You can read the problem and know what two or three numbers must be used, but can't go any further. Try asking yourself "am I trying to make the number bigger or break it down into its smaller parts". If you are trying to make the number bigger, use the LCM and if you're trying to break it down into smaller parts, use the GCF. This is not a guarantee for success, but it should be helpful.

The Grade 5 book pages shown simply demonstrate the basic computation of the greatest common factor. In many texts, you will see LCM and GCF discussed as stand alone topics as well as when working with fractions. Be prepared to see these topics often when you are teaching.

The *Focus on Skills* page allows you to practice finding answers using any method. However, if your method is to list factors and multiples, you might want to find a more expedient method. In *Focus on Concepts*, why some of those methods work is explored more in depth. It also looks at some relatively prime numbers. Word problems are included as application problems.

# Greatest Common Factor

**You Will Learn**
how to find the greatest common factor

**Vocabulary**

**common factor**
a number that is a factor of two or more different numbers

**greatest common factor (GCF)**
the greatest number that is a factor of each of two or more numbers

**Remember**
A number is a factor of another number if it divides the number with a remainder of zero.

### Learn

A group of 18 scientists and 24 photographers are scuba diving to study a coral reef. The group must split up equally into same-sized teams and take turns diving. There should be as many teams as possible.

**One Way**

Same-sized teams of scientists
Factors of 18: 1, 2, 3, ⑥, 9, 18

Same-sized teams of photographers
Factors of 24: 1, 2, 3, 4, ⑥, 8, 12, 24

GCF is 6
$18 \div 6 = 3$　$24 \div 6 = 4$

**Another Way**

Teams of scientists　Teams of photographers
Factors of 18　Factors of 24

9　1　4
18　2　
　3　12
6　8　24
Common factors

GCF is 6
$18 \div 6 = 3$　$24 \div 6 = 4$

There should be 6 teams of 3 scientists and 4 photographers.

### Talk About It

What number is a factor of every whole number? Explain.

### Check

Find the common factors and the greatest common factor for each pair.

**1.** 6 and 8　　**2.** 9 and 18　　**3.** 12 and 30　　**4.** 7 and 8

**5.** Find the greatest common factor of 12, 24, and 36.

**6. Reasoning** If 15 is a factor of a number, what other numbers must also be factors?

## Practice

### Skills and Reasoning

Find the common factors and the greatest common factor for each pair.

**7.** 8 and 12          **8.** 4 and 6          **9.** 10 and 15          **10.** 6 and 18

**11.** 9 and 3          **12.** 18 and 27        **13.** 16 and 20         **14.** 20 and 30

**15.** 12 and 24        **16.** 9 and 30         **17.** 10 and 45         **18.** 16 and 24

**19.** Find the greatest common factor of 8, 16, and 24.

**20.** Find the factors and the common factors of 6 and 15.

**21.** Could 10 be the greatest common factor of 35 and 40? Explain.

**22.** The common factors of two numbers are 1 and 3. The two numbers could be 9 and 21 or 3 and 6. Explain how.

### Problem Solving and Applications

**23. Patterns** A scientist has 36 photos of reef fish that she wants to arrange in equal rows and columns to frame. How many photo arrangements can be made?

**24. Science** A scientist is setting up some study tanks. She has collected 12 identical fish and 15 identical plants. She wants all tanks to be alike and contain as many fish and plants as possible. What is the greatest number of tanks she can set up?

### Mixed Review and Test Prep

Copy and complete.

**25.** $\frac{2}{4} = \frac{4}{\blacksquare}$     **26.** $\frac{1}{\blacksquare} = \frac{2}{12}$     **27.** $\frac{4}{12} = \frac{\blacksquare}{3}$     **28.** $\frac{2}{5} = \frac{\blacksquare}{10}$     **29.** $\frac{\blacksquare}{3} = \frac{8}{12}$

Find equivalent fractions.

**30.** $\frac{20}{40}$     **31.** $\frac{8}{12}$     **32.** $\frac{8}{16}$     **33.** $\frac{2}{8}$     **34.** $\frac{6}{16}$

**Algebra Readiness** Copy and complete.

**35.** $5 \times n = 30$     **36.** $3 \times n = 12$     **37.** $n \times 4 = 20$     **38.** $n \times 3 = 21$

**39.** Which number correctly completes the number sentence $\frac{3}{7} = \frac{\blacksquare}{14}$?

   ⓐ 7       ⓑ 10       ⓒ 6       ⓓ 11

**FOCUS ON SKILLS**
**LCM and GCF**

Name _____

1.    LCM (15, 25)                    2.    GCF (15, 25)

3.    LCM (21, 49)                    4.    GCF (21, 49)

5.    LCM (18, 32)                    6.    LCM (18, 32)

7.    LCM (22, 35)                    8.    GCF (22, 35)

9.    LCM (24, 36)                    10.   GCF (24, 36)

11.   LCM (117, 273)                  12.   GCF (117, 273)

13.   LCM (10, 15, 20)                14.   GCF (10, 15, 20)

15.   LCM (12, 14, 18)                16.   GCF (12, 14, 18)

**FOCUS ON CONCEPTS**                     Name _____
**LCM and GCF**

In the following Venn diagram, A represents the prime factorization of 15 and B represents the prime factorizatiom of 25.  It is used to show that the least common multiple is 5.  It also shows that the greatest common factor is 75.

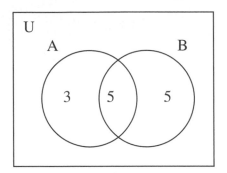

1.    a.    Use this method to find the LCM and GCF for 24 and 36.

      b.    Discuss the advantages and disadvantages of this method.

2.    a.    Find LCM (13, 27) and GCF (13, 27).

      b.    Do you see anything interesting about those numbers?

      c.    Look up relatively prime in your text or some other source and explain what it means in everyday language.

      d.    When two numbers are relatively prime, how does it effect the LCM  and GCF of those numbers?

Each of the following problems can be solved using an LCM or a GCF.  Not only should you solve the problem, but also explain your strategy for solving.

3.　　Janna goes to the laundry room in the dorm every 8 days.  Kevin goes to the laundry room in the dorm every 10 days.  If they meet in the laundry room while doing their laundry, how long will it be before they meet again in the laundry room?

4.　　Matt drinks bottled water.  He has a 24 ounce bottle that he refills all through the day from gallon jugs.  Each jug has 128 ounces.  When will a one-gallon jug run out at the same time he finishes filling a bottle?

5.　　The school has a case of 144 candy bars and a case of 24 sodas.  If these are divided evenly among the students, how many students will get candy and soda?  How many candy bars and sodas will each person get?

6.　　For a family picnic, Dana's mom buys a box of 12 forks, a box of 18 spoons, and a box of 24 knives.  How many boxes of each will she have to buy in order to have the same number of forks, spoons, and knives?

## FRACTIONS

One of the biggest surprises for students and some faculty is realizing when fractions first appear. The concept of half is in the Kindergarten book. Fractions are studied at all levels you are going to teach. The basic foundations are crucial for you to understand.

Some college students have poor fraction skills. Many calculators will perform fraction operations, so there seems to be no need for real understanding as long as a correct answer is produced. Part of the problem lies not just in operations with fractions, but with the basic understanding of what fractions are. It is your job not just to know how to manipulate fractions, but to fully understand the concepts of fractions and how important they are in what you will be teaching.

Your MET book probably has several approaches for knowing what fractions are. These different approaches to the same fraction are not meant to confuse you with several different representations or strange terminology. These alternative strategies are designed to give you a more thorough understanding of the concept of fractions.

The part-to-whole approach is a common way for students at all levels to think about fractions. Many times this is presented as a picture of a whole with a part of the picture shaded. This part-to-whole meaning allows for an easier discussion of equivalent fractions. The division approach is usually not the first introduction to fractions, but is useful when you start to move from fractions to decimals.

With the beginnings of fractions being seen in the Kindergarten book, you can see how important it is to fully understand fractions and be able to explain them. A large portion of the Grades 3, 4, and 5 books are devoted to fractions. The concept is everywhere, no matter what grade level you plan to teach. The text pages shown are from the Grade 2 book. It is interesting to note that many times fractions are discussed using a set model and Venn diagrams.

One topic often seen in fraction sections of texts is equivalent fractions. They are an essential tool for working with many types of fraction problems. You will have to understand equivalent fractions thoroughly in order to do computations with fractions and to teach your students to do the same.

*Focus on Skills* emphasizes the part-to-whole fraction model and equivalent fractions, while *Focus on Concepts* asks for writing and explanations. Again, written communication in mathematics is an essential skill. You cannot simply be able to talk about a subject, but you must also be able to write your ideas clearly and effectively.

Name _____

# Fractions

**Learn**

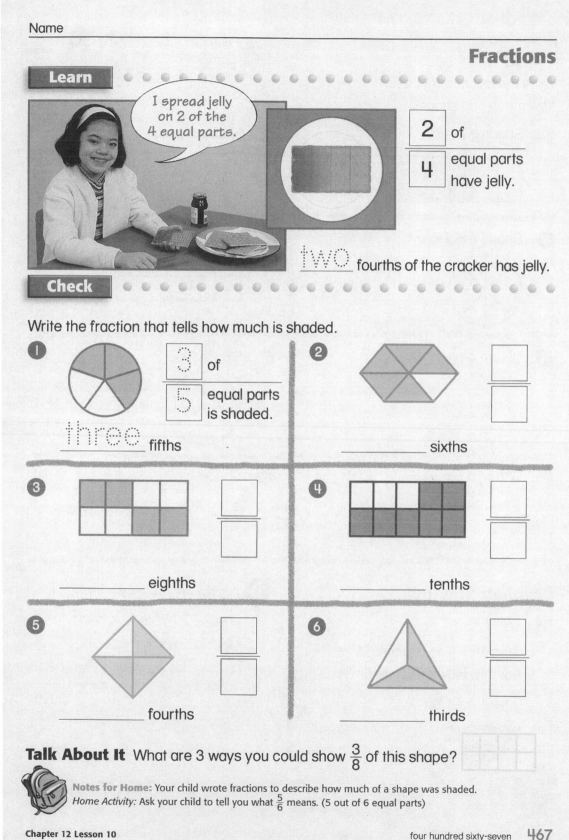

I spread jelly on 2 of the 4 equal parts.

$\dfrac{2}{4}$ of equal parts have jelly.

two fourths of the cracker has jelly.

**Check**

Write the fraction that tells how much is shaded.

1. $\dfrac{3}{5}$ of equal parts is shaded.

   three fifths

2. _____ sixths

3. _____ eighths

4. _____ tenths

5. _____ fourths

6. _____ thirds

**Talk About It** What are 3 ways you could show $\frac{3}{8}$ of this shape?

**Notes for Home:** Your child wrote fractions to describe how much of a shape was shaded.
*Home Activity:* Ask your child to tell you what $\frac{5}{6}$ means. (5 out of 6 equal parts)

**Practice** ● ● ● ● ● ● ● ● ● ● ● ● ●

Write the fraction that tells
how much you shaded.

**7** Shade 3 parts.

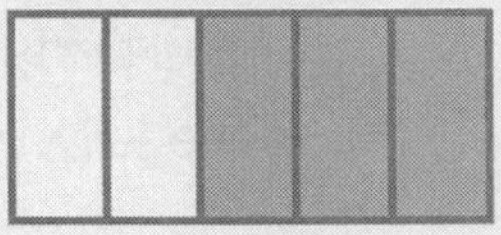

$\dfrac{3}{5}$

**8** Shade 2 parts.

**9** Shade 6 parts.

**10** Shade 9 parts.

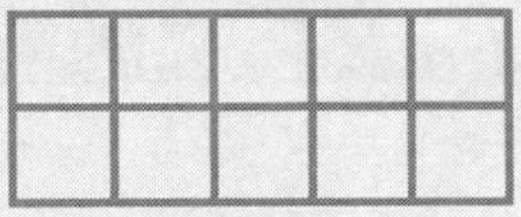

**11** Shade 1 part.

**12** Shade 2 parts.

**13** Shade 3 parts.

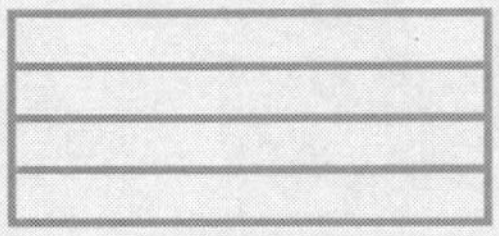

**14** Shade 1 part.

PRACTICE

## Problem Solving

**15** Solve.

You have $\dfrac{1}{3}$ of a granola
bar left. How much did
you already eat?

**Write your own** problem
about a fraction of a pizza.
Have a friend solve it.

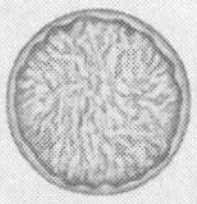

_____

_____

_____

**Notes for Home:** Your child practiced writing fractions. _Home Activity:_ Ask your child to fold a paper
into 4 equal parts, shade some parts, and name the fraction that describes how much of the paper
was shaded.

For additional practice, see Skills Practice Bank, page 538, Set 2.

**FOCUS ON SKILLS**                                    Name _____
**Fractions**

For each figure below, write two fractions:  one to indicate what part of the figure is
shaded and one to indicate what part of the figure is unshaded.

1.                                                              2.

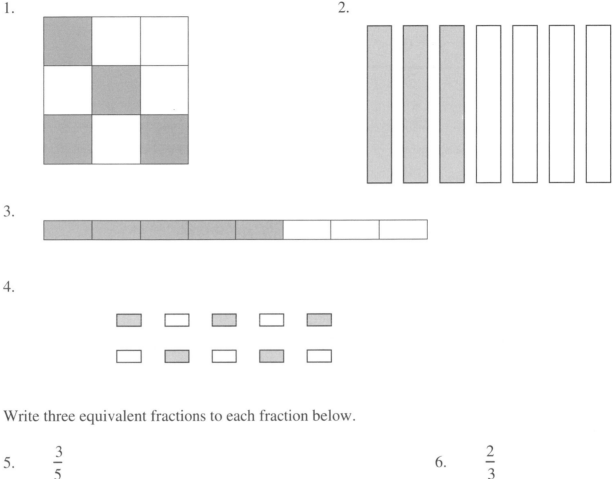

3.

4.

Write three equivalent fractions to each fraction below.

5.    $\dfrac{3}{5}$                                           6.    $\dfrac{2}{3}$

7.    $\dfrac{11}{10}$                                         8.    $\dfrac{10}{13}$

Fill in the missing number to make the two fractions in each set equivalent.

9.    $\dfrac{3}{7} = \dfrac{12}{\quad}$          10.    $\dfrac{\quad}{8} = \dfrac{25}{40}$          11.    $\dfrac{5}{6} = \dfrac{\quad}{30}$

12.    Suppose you need to drill a 5/8 inch hole, but the set of drill bits is measured in
       sixteenths of an inch.  What size drill bit should you use?

**FOCUS ON CONCEPTS**          **Name** _____
**Fractions**

Many questions on this sheet are related to the Focus on Skills page.  If you have not already completed those problems, you will find it helpful before beginning this sheet.

1.     What did you notice about the relationship between the fraction for the shaded region and the fraction for the unshaded region?

2.     Will this always be the case?  Explain.

3.     How is this information helpful in the study of fractions?

4.     Explain the process you used for creating equivalent fractions.

5.     Is there another way other than the method you use?  Explain it.

6.     Give two ways you can tell if two fractions are equivalent.

7.     Explain in a few sentences why equivalent fractions are important in later work with fractions.

## ADDITION AND SUBTRACTION OF FRACTIONS

These topics are seen at all levels beginning with grade 3. The sample pages are from a Grade 6 book. There are numerous pages in the school texts devoted to addition and subtraction of fractions. We chose these pages because they emphasize equivalent fractions and least common denominators as the method for solving the problems. While you may be more comfortable with another method, this is the preferred method in school texts.

Some students want to use $\frac{a}{b} \pm \frac{c}{d} = \frac{ad \pm bc}{bd}$ to find a solution, when many times it involves more work than simply finding the common denominator. You may find that many MET books use this approach for addition and subtraction of fractions. If you have never had much success with fractions, this method does provide some security; it always works. However, as future teachers you must get past this insecurity in order to be an effective teacher. The pages from the school text demonstrate the need for knowing and learning the LCD method.

Some of our students are still reluctant to use LCD's and equivalent fractions because they are not comfortable with the method. That is why we give a lot of skill-based problems for practice. We try to raise our students' comfort level through these types of problems. We do this by using problems like you'll see in the *Focus on Skills* page. These are all straightforward computation problems. An example is given for you and you are expected to follow that format to get the answers.

The *Focus on Concepts* problems highlight the $\frac{a}{b} \pm \frac{c}{d} = \frac{ad \pm bc}{bd}$ approach that is found in many college textbooks. It reinforces the idea that using an LCD can save a lot of work in many cases. It also points out that this method is a nice tool when both the denominators are prime. There are also several word problems included. No matter how simple the fraction operation is, our students are still scared of word problems. Perhaps you are too. These problems are not included because they are difficult, but to try and raise your comfort level in working with word problems.

Many MET textbooks are including higher order thinking problems and problems that focus on concepts. Your text should have an ample supply of these. The activities here are for additional practice. Our students need help with both skills and critical thinking and this is what we use with them.

# Adding and Subtracting Fractions with Like Denominators

**You Will Learn**
- to add and subtract fractions with like denominators

**Vocabulary**
like denominators

**Learn**

David lives near the Museum of Life and Science, which has a very popular hands-on exhibit of the circulatory system. David is planning to see the entire museum in one week.

David is from Durham, North Carolina.

Two fractions with the same denominator have **like denominators**. When you add and subtract fractions with like denominators, the denominator acts as a label. It tells you what size pieces you're using. The numerators tell the number of pieces you add or subtract.

**Example 1**

By Tuesday, David had been through $\frac{2}{7}$ of the museum. By Saturday, he had been through another $\frac{4}{7}$ of the museum. How much of the museum has David seen so far?

Add: $\frac{2}{7} + \frac{4}{7}$

$\frac{2}{7} + \frac{4}{7} = \frac{2+4}{7}$    Add numerators only.

$\frac{2}{7} + \frac{4}{7} = \frac{6}{7}$    Denominators do not change.

**Remember**
Like denominators are also known as common denominators.

**Example 2**

Subtract: $\frac{4}{8} - \frac{1}{8}$

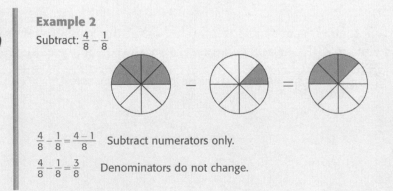

$\frac{4}{8} - \frac{1}{8} = \frac{4-1}{8}$    Subtract numerators only.

$\frac{4}{8} - \frac{1}{8} = \frac{3}{8}$    Denominators do not change.

You can add and subtract improper fractions with like denominators. Write the answer as a mixed number.

**Example 3**

In 1940, Charles Drew revolutionized the way doctors cared for patients by devising a blood bank plan for adequate storage of blood.

When an adult man donates blood to a blood bank, he donates about $\frac{1}{2}$ of a quart. The body of an average man contains about 5, or $\frac{10}{2}$, quarts of blood. How much blood is in his body after the donation?

$$\frac{10}{2} - \frac{1}{2} = \frac{10-1}{2}$$
$$= \frac{9}{2} = 4\frac{1}{2}$$

There are $\frac{9}{2}$, or $4\frac{1}{2}$ quarts, left in his body.

When adding and subtracting fractions, write the sum or difference in lowest terms.

**Example 4**

Add:  $\frac{5}{16} + \frac{3}{16}$

$$\frac{5}{16} + \frac{3}{16} = \frac{5+3}{16}$$
$$= \frac{8}{16} = \frac{2}{4} = \frac{1}{2}$$

$$\frac{5}{16} + \frac{3}{16} = \frac{1}{2}$$

**Example 5**

Subtract:  $\frac{7}{8} - \frac{1}{8}$

$$\frac{7}{8} - \frac{1}{8} = \frac{7-1}{8}$$
$$= \frac{6}{8} = \frac{3}{4}$$

$$\frac{7}{8} - \frac{1}{8} = \frac{3}{4}$$

**Talk About It**

1. When you add or subtract fractions with like denominators, why doesn't the denominator change?

2. What values can $n$ have to make the equation $\frac{3}{n} + \frac{5}{n} = \frac{8}{n}$ true?

**Check**

Simplify. Write each answer in lowest terms.

1. $\frac{3}{10} + \frac{4}{10}$    2. $\frac{5}{7} - \frac{3}{7}$    3. $\frac{8}{2} + \frac{9}{2}$    4. $\frac{4}{9} - \frac{4}{9}$    5. $\frac{3}{4} - \frac{2}{4}$

6. $\frac{5}{3} - \frac{2}{3}$    7. $\frac{1}{6} + \frac{7}{6}$    8. $\frac{1}{5} + \frac{1}{5}$    9 $\frac{6}{8} - \frac{2}{8}$    10. $\frac{11}{12} + \frac{10}{12}$

11. **Reasoning**  How is adding fractions with like denominators similar to subtracting fractions with like denominators?

**Practice** • • • • • • • • • • • • • • • • • • • • • • • • • • • • • • • •

## Skills and Reasoning

Simplify. Write each answer in lowest terms.

**12.** $\frac{3}{5} + \frac{1}{5}$     **13.** $\frac{9}{10} - \frac{8}{10}$     **14.** $\frac{7}{8} + \frac{5}{8}$     **15.** $\frac{4}{3} + \frac{2}{3}$     **16.** $\frac{12}{5} - \frac{6}{5}$

**17.** $\frac{4}{3} - \frac{3}{3}$     **18.** $\frac{98}{10} + \frac{2}{10}$     **19.** $\frac{3}{4} - \frac{1}{4}$     **20.** $\frac{4}{11} + \frac{3}{11}$     **21.** $\frac{12}{18} - \frac{9}{18}$

**22.** $\frac{15}{19} + \frac{5}{19}$     **23.** $\frac{7}{9} - \frac{3}{9}$     **24.** $\frac{6}{8} - \frac{4}{8}$     **25.** $\frac{5}{13} + \frac{1}{13}$     **26.** $\frac{34}{12} - \frac{30}{12}$

State whether the answer is greater than, less than, or equal to 1.

**27.** $\frac{7}{9} + \frac{2}{9}$     **28.** $\frac{1}{2} + \frac{3}{2}$     **29.** $\frac{2}{7} + \frac{6}{7}$     **30.** $\frac{3}{4} - \frac{2}{4}$

**31.** $\frac{9}{5} - \frac{4}{5}$     **32.** $\frac{7}{12} + \frac{7}{12}$     **33.** $\frac{1}{10} - \frac{1}{10}$     **34.** $\frac{16}{13} + \frac{4}{13}$

Tillie's volleyball team had a picnic. Team members brought food or games. The bar graph represents the players who brought an item of food. Use the graph for **35–41.**

**35.** What fraction of the students shown in the graph brought fruit or drinks?

**36.** What is the difference between the fraction of students who brought drinks and the fraction who brought fruit?

**37.** What fraction of the students shown in the graph brought fruit, drinks, or salad?

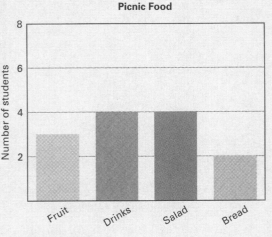

**38.** If 15 students attended the picnic, how many did not bring fruit, drinks, salad, or bread?

**39.** What is the difference between the fraction of students who brought fruit or bread and the fraction who brought salad?

**40.** Two more students attended the picnic at the last minute and both brought bread. What would the new fraction of students bringing bread be?

**41.** **Critical Thinking** Which two foods represent over $\frac{1}{2}$ of the foods brought to the picnic?

**FOCUS ON SKILLS**                    Name _____
**Addition and Subtraction of Fractions**

Perform the following addition and subtraction problems.  Use the problem below as an example.  Show all your work.  Your answer must be in simplest form.

$$\frac{1}{3}+\frac{2}{5}=\frac{5}{15}+\frac{6}{15}=\frac{11}{15}$$

1.    $\frac{3}{7}+\frac{1}{3}=$                    2.    $\frac{8}{9}+\frac{1}{12}=$

3.    $\frac{3}{8}+\frac{27}{40}=$                    4.    $\frac{1}{4}+\frac{5}{6}=$

5.    $\frac{7}{10}+\frac{11}{16}=$                    6.    $\frac{7}{8}+\frac{5}{12}=$

7.    $\frac{19}{24}-\frac{5}{8}=$                    8.    $\frac{13}{18}-\frac{8}{27}=$

9.    $\frac{5}{8}-\frac{3}{16}=$                    10.    $\frac{3}{5}-\frac{9}{20}=$

11.    $\frac{7}{12}-\frac{3}{10}=$                    12.    $\frac{23}{30}-\frac{11}{20}=$

**FOCUS ON CONCEPTS**                    Name _____
**Addition and Subtraction of Fractions**

1.     About eleven-twelfths of a garden is planted with tomatoes, one-eighteenth with onions, and the rest with peppers.  What part of the garden is planted with peppers?

2.     Donna ate one-fourth of a pizza and Don ate one-third of it.  What fraction of the pizza did they eat?

3.     In making a budget, a person should spend about one-third of his or her salary on rent or housing, should put about one-tenth into a savings account, and should plan to have about one-third taken out in taxes.  What fraction of a person's salary is then left for everything else?

4.     A runner jogs two-fifths of a mile on Monday, Wednesday, and Friday and two-thirds of a mile on Tuesday and Thursday.  What is the total distance for this runner for the week?

5.     There is a method for adding and subtracting fractions that does not require you to obtain a least common denominator in order to solve the problem.  This method says:

$$\frac{a}{b} \pm \frac{c}{d} = \frac{ad \pm bc}{bd}$$

An illustration of this is:

$$\frac{2}{3} + \frac{4}{5} = \frac{(2)(5)+(3)(4)}{(3)(5)} = \frac{10+12}{15} = \frac{22}{15} = 1\frac{7}{15}$$

However, when considering the problem $\frac{23}{30} - \frac{11}{20}$, would you think this is the most appropriate method for solution?

6.    Explain the advantages and disadvantages of this method using specific examples in your explanation.

7.    A student wants to know why $\frac{2}{3}+\frac{1}{5}$ is not $\frac{3}{8}$.  How do you answer this student's question?  Discuss all the possible ways you can convince this student of the correct answer.

8.    Give three examples of how addition or subtraction of fractions is applicable to a person's everyday life.  Explain your choices.

## MULTIPLICATION AND DIVISION OF FRACTIONS

Multiplication with fractions is straightforward and considered to be much easier than addition or subtraction.  Division with fractions is simply one additional step, and then it becomes a multiplication problem.  Because many view multiplying and dividing fractions as easier tasks, they teach it before adding and subtracting.  Your MET book or the text you teach from may take this approach.  There are both pros and cons to this method.

By looking at multiplication and division first, you ease students into operations with fractions in a less threatening environment.  The computations are fairly straightforward and not very difficult, especially in the beginning.  There are no common denominators to deal with.  The only snags may occur when simplifying an answer (should this be done to the final answer or along the way).  Still, the ease of the calculations is the main reason multiplication and division of fractions is taught before addition and subtraction.

However, if students become too accustomed to doing a multiplication problem as $\frac{a}{b} \times \frac{c}{d} = \frac{ac}{bd}$, it is possible that they will be tempted to do their addition and subtraction problems in the same manner.  Also, since multiplication is often taught as repeated addition, it is necessary to have the addition skills before beginning with multiplication.  The same argument holds true for division as repeated subtraction.  It is really a toss-up as to which should be taught first.  We have tried both methods with neither coming out a clear winner.

Multiplication with fractions is not very difficult to explain in terms of the algorithm or the model.  By using the same type of array model as with whole number multiplication, you can see a relationship back to whole numbers as well as to the algorithm.

Relating the "invert and multiply" algorithm to a picture model is somewhat more difficult.  I usually start with something as simple as the following:  There are 6 cookies that you want to divide into halves.  If I draw a line down the middle of those 6 cookies, I now have 12 pieces.  So $6 \div \frac{1}{2} = 6 \times \frac{2}{1} = 12$.  Another common approach to division is with common denominators.  This is seen in the *Focus on Concepts* page.  The *Focus on Skills* page simply asks you to perform operations based on the sample given.

**Chapter 9**
**Lesson**
**4**

# Exploring Multiplication of Fractions by Fractions

**Problem Solving Connection**
- Look for a Pattern
- Use Objects/ Act It Out

**Materials**
- paper squares
- blue, red pencils

 **Explore** • • • • • • • • • • • • • • • • • • • • • • • •

How can you find a fraction of a fraction?

## Work Together

You can find a fraction of a fraction by paper folding and shading.

**Did You Know?**
Origami is the Japanese art of paper folding. It dates back to at least 1682.

1. Use paper folding to find $\frac{1}{2}$ of $\frac{1}{4}$.
   a. Fold a square of paper vertically down the center. What fraction of the whole piece is each section?
   b. Fold the paper once more vertically to get four equal sections. What fraction is each section?
   c. Now fold the paper in half horizontally. How many sections are there? What fraction of the square is each small section?

2. Use your folded paper.
   a. Shade $\frac{3}{4}$ red vertically.    b. Shade $\frac{1}{2}$ blue horizontally.
   c. The purple section shows $\frac{1}{2}$ of $\frac{3}{4}$. What is $\frac{1}{2}$ of $\frac{3}{4}$?

3. Fold a new square of paper to find $\frac{1}{2}$ of $\frac{1}{2}$.

4. Use paper folds to find each answer.
   a. $\frac{3}{4}$ of $\frac{1}{8}$    b. $\frac{1}{4}$ of $\frac{7}{8}$    c. $\frac{3}{4}$ of $\frac{3}{8}$    d. $\frac{1}{4}$ of $\frac{3}{8}$

## Talk About It

5. In the illustration for **1c** above, what fraction of the whole do two sections make?

6. Explain how you could use paper folding to find $\frac{1}{3}$ of $\frac{1}{5}$.

**Connect** • • • • • • • • • • • • • • • • • • • • • • • • • • • • • • • • • • • • •

$\frac{3}{8} \times \frac{1}{2}$ means $\frac{3}{8}$ of $\frac{1}{2}$. The drawings show $\frac{3}{8} \times \frac{1}{2} = \frac{3}{16}$.

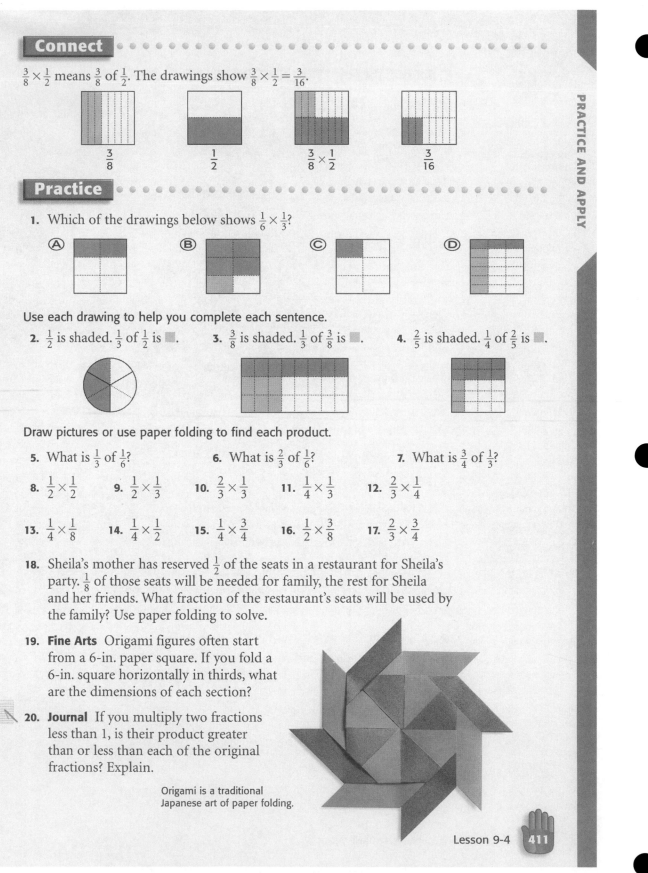

$\frac{3}{8}$       $\frac{1}{2}$       $\frac{3}{8} \times \frac{1}{2}$       $\frac{3}{16}$

**Practice** • • • • • • • • • • • • • • • • • • • • • • • • • • • • • • • • • • • • •

1. Which of the drawings below shows $\frac{1}{6} \times \frac{1}{3}$?

Ⓐ      Ⓑ      Ⓒ      Ⓓ

Use each drawing to help you complete each sentence.

2. $\frac{1}{2}$ is shaded. $\frac{1}{3}$ of $\frac{1}{2}$ is ▮.

3. $\frac{3}{8}$ is shaded. $\frac{1}{3}$ of $\frac{3}{8}$ is ▮.

4. $\frac{2}{5}$ is shaded. $\frac{1}{4}$ of $\frac{2}{5}$ is ▮.

Draw pictures or use paper folding to find each product.

5. What is $\frac{1}{3}$ of $\frac{1}{6}$?

6. What is $\frac{2}{3}$ of $\frac{1}{6}$?

7. What is $\frac{3}{4}$ of $\frac{1}{3}$?

8. $\frac{1}{2} \times \frac{1}{2}$      9. $\frac{1}{2} \times \frac{1}{3}$      10. $\frac{2}{3} \times \frac{1}{3}$      11. $\frac{1}{4} \times \frac{1}{3}$      12. $\frac{2}{3} \times \frac{1}{4}$

13. $\frac{1}{4} \times \frac{1}{8}$      14. $\frac{1}{4} \times \frac{1}{2}$      15. $\frac{1}{4} \times \frac{3}{4}$      16. $\frac{1}{2} \times \frac{3}{8}$      17. $\frac{2}{3} \times \frac{3}{4}$

18. Sheila's mother has reserved $\frac{1}{2}$ of the seats in a restaurant for Sheila's party. $\frac{1}{8}$ of those seats will be needed for family, the rest for Sheila and her friends. What fraction of the restaurant's seats will be used by the family? Use paper folding to solve.

19. **Fine Arts** Origami figures often start from a 6-in. paper square. If you fold a 6-in. square horizontally in thirds, what are the dimensions of each section?

20. **Journal** If you multiply two fractions less than 1, is their product greater than or less than each of the original fractions? Explain.

Origami is a traditional Japanese art of paper folding.

**FOCUS ON SKILLS**                    Name _____

**Multiplication and Division of Fractions**

For problems 1 – 6, multiply the fractions given.  You can either multiply first and simplify later or simplify first and then multiply.  Whichever method you find best for you should be the one you choose.

1. $\dfrac{3}{5} \times \dfrac{4}{5} =$

2. $\dfrac{1}{3} \times \dfrac{3}{7} =$

3. $\dfrac{2}{5} \times \dfrac{10}{13} =$

4. $\dfrac{3}{7} \times \dfrac{14}{27} =$

5. $\dfrac{7}{9} \times \dfrac{18}{49} \times \dfrac{3}{4} =$

6. $\dfrac{19}{20} \times \dfrac{5}{6} \times \dfrac{2}{3} =$

For each division problem below, make sure you show every step of your work.

7. $\dfrac{1}{2} \div \dfrac{2}{3} =$

8. $\dfrac{2}{5} \div \dfrac{9}{2} =$

9. $\dfrac{3}{7} \div \dfrac{3}{8} =$

10. $\dfrac{4}{9} \div \dfrac{2}{3} =$

11. $\dfrac{6}{7} \div \dfrac{3}{5} =$

12. $\dfrac{4}{5} \div \dfrac{2}{5} =$

**FOCUS ON CONCEPTS**                    Name _____
**Multiplication and Division of Fractions**

1.   When you work the problem $\frac{3}{5} \times \frac{4}{5}$ you get an answer of $\frac{12}{25}$ which cannot be simplified.  What does this tell you about this problem and about other problems like it?

2.   Draw an array model for $\frac{3}{5} \times \frac{4}{5}$ and explain what it means.

Division problems can be computed using a common denominator as in the following example:

$$\frac{4}{5} \div \frac{2}{5} = 4 \div 2 = 2$$

which is verifies using the traditional "invert and multiply" algorithm, by using repeated subtraction, or drawing a picture of the situation.

2.   List two advantages to this method over the traditional algorithm.

3.   List two disadvantages to this method over the traditional algorithm.

4.   Use this method to solve $\frac{5}{7} \div \frac{2}{7}$.

5.   Explain the common denominator method used in #4 in your own words.

6.   As mentioned in the *Focus on Skills* sheet, when doing multiplication, you can either multiply first and then simplify or simplify first and then multiply.  Which method do you like best?  Why?  Describe the down side of your less preferred method.

# MIXED NUMBERS AND IMPROPER FRACTIONS

Mixed numbers should receive more attention than they do. Often they are introduced quickly, converted to improper fractions, and then the rest of the time is spent on operations with mixed numbers. It is the operations with mixed numbers that so many students have trouble with. By spending more time on the foundations of mixed numbers and some additional practice with the concepts, some of the confusion with operations may be eliminated.

What is a mixed number? It is a mixture or combination of a whole number with a fraction. By changing that whole number to a fraction, it is clear that the mixed number can be written in fraction form, called an improper fraction. So a mixed number can be written as an improper fraction and vice versa, which is why the two topics are interrelated.

However, the way a mixed number is written is often confusing. The mixed number $4\frac{2}{3}$ is actually representing the quantity 4 and $\frac{2}{3}$, which is how it is spoken. What this number is telling us to do is to combine the 4 with the $\frac{2}{3}$ through the process of addition.

Most students tend to interpret it is multiplication, or just simply don't understand what it means at all. Carefully drawing pictures of mixed numbers can help with the understanding.

The term "improper fractions" sounds like there is something wrong with these fractions. All is really means is that the numerator is larger than the denominator and that is not the typical way we write fractions. In later mathematics courses you often find improper fractions as a very acceptable way of presenting an answer. But in the school texts, improper fractions are always changed to mixed numbers. Therefore, it is a necessary skill for you to understand.

The conversion from improper fraction to mixed number is simply a division problem with a remainder that many students pick up on quickly. The conversion from mixed number to improper fraction follows a very traditional algorithm that tends to be confusing without some understanding of how it works. The *Focus on Concepts* page explores this algorithm. The *Focus on Skills* page gives you practice in converting improper fractions and mixed numbers.

The school text pages are from the Grade 4 book, but appear as early as the Grade 3 text. Mixed numbers and improper fractions are important school topics and must be understood, not just manipulated. It is your job to make sure you fully understand these topics and the underlying concepts behind them.

# Exploring Mixed Numbers

**Problem Solving
Connection**

■ Use Objects/
  Act It Out

■ Draw a Picture

**Materials**
fraction strips

**Vocabulary**
**mixed number**
a number that
has a whole
number and a
fractional part

**improper fraction**
a fraction in which
the numerator is
greater than or
equal to the
denominator

**Math Tip**
If two sets of
fraction strips are
the same length,
they show the
same amount.

**Explore** ● ● ● ● ● ● ● ● ● ● ● ● ● ● ● ● ● ● ● ● ● ● ● ● ● ● ● ● ● ●

A bowl of Brand D cereal has
$2\frac{1}{2}$ teaspoons of sugar. The
number $2\frac{1}{2}$ is a **mixed number**.
If you use a $\frac{1}{2}$ teaspoon to
show this amount of sugar,
you will fill the spoon five times.

The **improper fraction** $\frac{5}{2}$ also
shows this amount.

You can use fraction strips
and number lines to explore
mixed numbers and improper
fractions.

## Work Together

1. Use fraction strips to show $1\frac{3}{4}$.

   **a.** How many $\frac{1}{4}$ strips are in 1?

   **b.** How many $\frac{1}{4}$ strips are in $\frac{3}{4}$?

   **c.** How many $\frac{1}{4}$ strips are in $1\frac{3}{4}$?

   **d.** What improper fraction shows
   how many $\frac{1}{4}$ strips are in $1\frac{3}{4}$?

2. Use fraction strips to show $1\frac{1}{8}$ on
   a number line.

   **a.** How many $\frac{1}{8}$ strips are in $1\frac{1}{8}$?

   **b.** Draw a number line below the
   fraction strips. Label the number line.

### Talk About It

1. How many $\frac{1}{8}$ fraction strips are in $2\frac{1}{8}$.

2. On a number line does $\frac{8}{3}$ come before $2\frac{1}{3}$?
   Explain.

PRACTICE AND APPLY

## Connect • • • • • • • • • • • • • • • • • • • • • • • • • • • • • • • • • • • • •

To write a mixed number as an improper fraction, break the whole number into fractional parts and add the parts.

For $1\frac{2}{3}$:

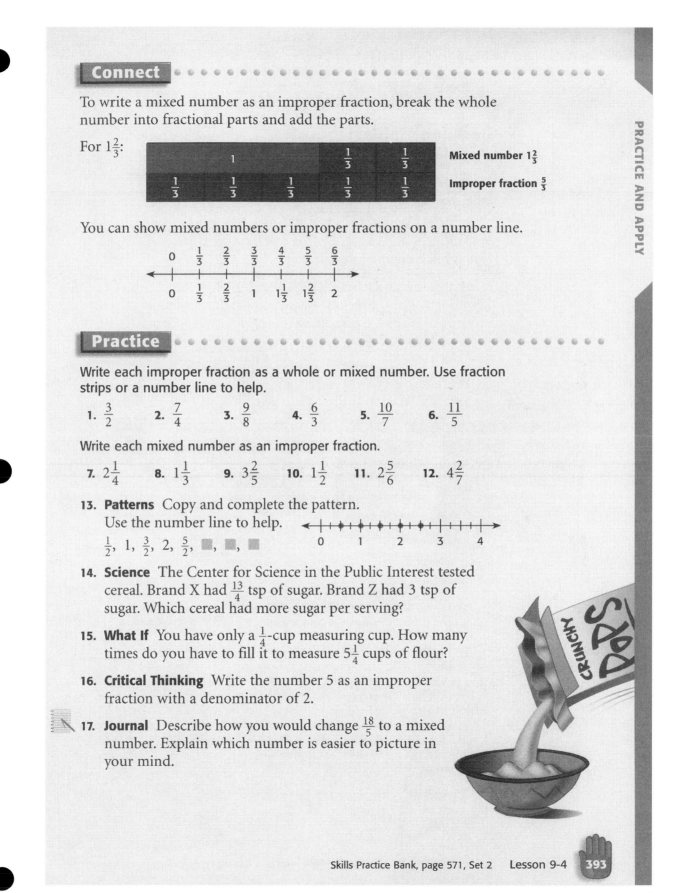

Mixed number $1\frac{2}{3}$

Improper fraction $\frac{5}{3}$

You can show mixed numbers or improper fractions on a number line.

## Practice • • • • • • • • • • • • • • • • • • • • • • • • • • • • • • • • • • • • •

Write each improper fraction as a whole or mixed number. Use fraction strips or a number line to help.

1. $\frac{3}{2}$    2. $\frac{7}{4}$    3. $\frac{9}{8}$    4. $\frac{6}{3}$    5. $\frac{10}{7}$    6. $\frac{11}{5}$

Write each mixed number as an improper fraction.

7. $2\frac{1}{4}$    8. $1\frac{1}{3}$    9. $3\frac{2}{5}$    10. $1\frac{1}{2}$    11. $2\frac{5}{6}$    12. $4\frac{2}{7}$

13. **Patterns** Copy and complete the pattern. Use the number line to help.

   $\frac{1}{2}$, 1, $\frac{3}{2}$, 2, $\frac{5}{2}$, ■, ■, ■

14. **Science** The Center for Science in the Public Interest tested cereal. Brand X had $\frac{13}{4}$ tsp of sugar. Brand Z had 3 tsp of sugar. Which cereal had more sugar per serving?

15. **What If** You have only a $\frac{1}{4}$-cup measuring cup. How many times do you have to fill it to measure $5\frac{1}{4}$ cups of flour?

16. **Critical Thinking** Write the number 5 as an improper fraction with a denominator of 2.

17. **Journal** Describe how you would change $\frac{18}{5}$ to a mixed number. Explain which number is easier to picture in your mind.

**FOCUS ON SKILLS**                          Name _____
**Mixed Numbers and Improper Fractions**

The mixed number $2\frac{3}{5}$ can be modeled by the picture below to show that $2\frac{3}{5} = \frac{13}{5}$.

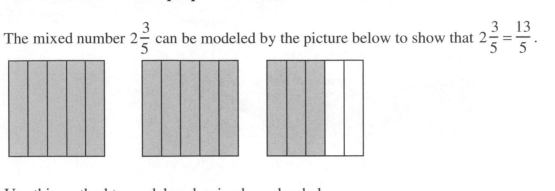

Use this method to model each mixed number below.

1.     $3\frac{1}{3}$                    2.     $1\frac{1}{2}$                    3.     $4\frac{2}{7}$

Convert each mixed number to an improper fraction.

4.     $3\frac{1}{3}$                    5.     $1\frac{1}{2}$                    6.     $4\frac{2}{7}$

Change each mixed number to an improper fraction.  Show all your work.

7.     $\frac{12}{5}$                    8.     $\frac{28}{3}$                    9.     $\frac{97}{15}$

**FOCUS ON CONCEPTS**                    **Name** _____
**Mixed Numbers and Improper Fractions**

In the *Focus on Skills* page, you saw a picture of $2\frac{3}{5}$ and how it relates to the improper

fraction $\frac{13}{5}$ which can also be obtained through the standard algorithm. Another way to

look at $2\frac{3}{5}$ is as $2 + \frac{3}{5}$ which is really $\frac{10}{5} + \frac{3}{5} = \frac{13}{5}$.

1.    Describe how this method is directly related to the standard algorithm.

2.    Use this method to convert $1\frac{4}{7}$ to an improper fraction.

3.    Use this method to convert $2\frac{3}{4}$ to an improper fraction.

4.    Use the standard algorithm to convert $1\frac{4}{7}$ to an improper fraction.

5.    Use the standard algorithm to convert $2\frac{3}{4}$ to an improper fraction.

6.    Explain the standard algorithm to a student in language they will understand.

7.    Explain the addition method to a student in language they will understand.

## MIXED NUMBER OPERATIONS

Addition and subtraction with mixed numbers are areas where all students, elementary, middle, high school and college, have numerous difficulties. Many of these difficulties lie not just in the operations themselves, but in the underlying problems with mixed numbers. With a better understanding of mixed numbers, some of these difficulties can be cleared up. Multiplication and division are not as misunderstood, but some problems do exist.

Multiplication and division problems are completed by changing the mixed number to an improper fraction, performing the operation, and then changing the answer back into mixed number form. It is a relatively simple process, yet some students have a deep misconception. The *Focus on Concepts* page demonstrates where some of the problems lie when working with multiplication and division.

The majority of problems students have are with addition and subtraction. Again, the operations of addition and subtraction are not the real problem. It is performing those operations on mixed numbers with the regrouping that is often necessary that poses the difficulty.

In recent years, an alternative approach to working with mixed number addition and subtraction has become prevalent. Many teachers are teaching their students to work with addition and subtraction like they do multiplication and division, by changing everything to an improper fraction before performing the operation. While this method will work, it is often cumbersome, and not the method seen in school texts.

The school text pages are from the Grade 5 book. It shows subtraction of mixed numbers, where the majority of mixed number problems occur. Note that the regrouping method is used. The regrouping idea is not just for addition and subtraction of whole numbers. It appears any time there is a need to "borrow" from another number. These pages not only show the importance of regrouping, but that it is the method taught in schools and you must have a complete understanding of these operations when you begin teaching.

The *Focus on Skills* page is a collection of problems with all four operations. It is expected that you will show all your work and ALL steps involved. These steps are not intended for you to simply memorize a way of doing things, but for you to have a better understanding of working the problems.

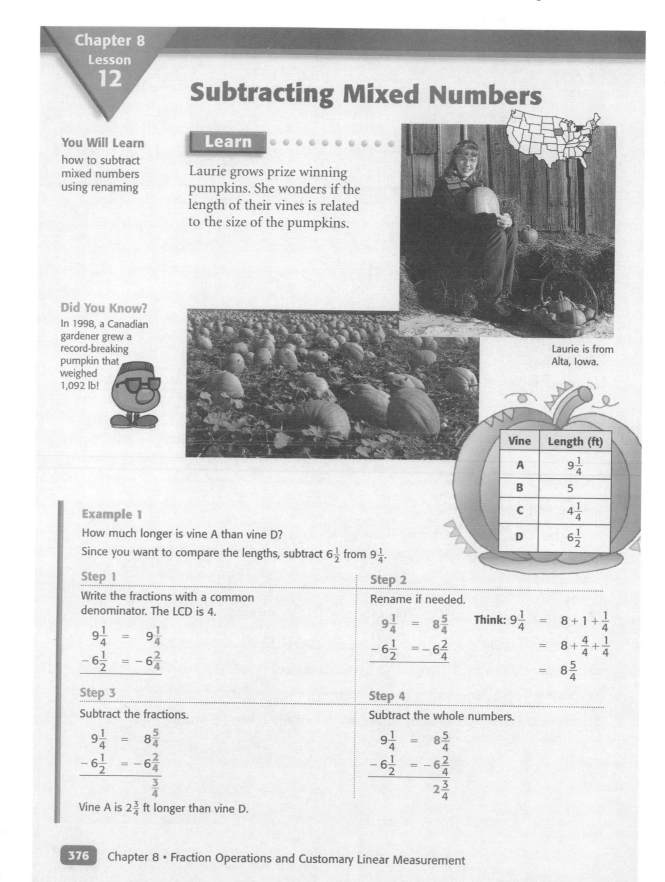

**Chapter 8**
**Lesson**
**12**

# Subtracting Mixed Numbers

**You Will Learn**
how to subtract
mixed numbers
using renaming

**Learn** • • • • • • • • • • •

Laurie grows prize winning
pumpkins. She wonders if the
length of their vines is related
to the size of the pumpkins.

**Did You Know?**
In 1998, a Canadian
gardener grew a
record-breaking
pumpkin that
weighed
1,092 lb!

Laurie is from
Alta, Iowa.

| Vine | Length (ft) |
|------|-------------|
| A | $9\frac{1}{4}$ |
| B | 5 |
| C | $4\frac{1}{4}$ |
| D | $6\frac{1}{2}$ |

### Example 1

How much longer is vine A than vine D?

Since you want to compare the lengths, subtract $6\frac{1}{2}$ from $9\frac{1}{4}$.

**Step 1**

Write the fractions with a common
denominator. The LCD is 4.

$$9\frac{1}{4} = 9\frac{1}{4}$$
$$-6\frac{1}{2} = -6\frac{2}{4}$$

**Step 2**

Rename if needed.

$$9\frac{1}{4} = 8\frac{5}{4}$$
$$-6\frac{1}{2} = -6\frac{2}{4}$$

**Think:** $9\frac{1}{4} = 8 + 1 + \frac{1}{4}$
$= 8 + \frac{4}{4} + \frac{1}{4}$
$= 8\frac{5}{4}$

**Step 3**

Subtract the fractions.

$$9\frac{1}{4} = 8\frac{5}{4}$$
$$-6\frac{1}{2} = -6\frac{2}{4}$$
$$\frac{3}{4}$$

**Step 4**

Subtract the whole numbers.

$$9\frac{1}{4} = 8\frac{5}{4}$$
$$-6\frac{1}{2} = -6\frac{2}{4}$$
$$2\frac{3}{4}$$

Vine A is $2\frac{3}{4}$ ft longer than vine D.

**Example 2**

How much longer is vine B than vine C?

$$
\begin{array}{r}
5 = 4\frac{4}{4} \\
-4\frac{1}{4} = -4\frac{1}{4} \\
\hline
\frac{3}{4}
\end{array}
$$

Vine B is $\frac{3}{4}$ ft longer than vine C.

**Example 3**

Another vine is $3\frac{2}{3}$ ft long. How much longer is vine C?

$$
\begin{array}{r}
4\frac{1}{4} = 4\frac{3}{12} = 3\frac{15}{12} \\
-3\frac{2}{3} = -3\frac{8}{12} = -3\frac{8}{12} \\
\hline
\frac{7}{12}
\end{array}
$$

Vine C is $\frac{7}{12}$ ft longer.

## Talk About It

1. In Example 1, how is renaming used to write $9\frac{1}{4}$ as $8\frac{5}{4}$?

2. In Example 2, how is renaming used to subtract $4\frac{1}{4}$ from 5?

## Check

**Copy and complete. Simplify.**

1.  $$
\begin{array}{r}
4\frac{1}{4} = 4\frac{3}{12} = 3\frac{\blacksquare}{12} \\
-2\frac{1}{3} = -2\frac{4}{12} = -2\frac{4}{12} \\
\hline
1\frac{\blacksquare}{12}
\end{array}
$$

2.  $$
\begin{array}{r}
5 = \blacksquare\frac{8}{8} \\
-2\frac{3}{8} = -2\frac{3}{8} \\
\hline
\blacksquare\frac{5}{8}
\end{array}
$$

3.  $$
\begin{array}{r}
3\frac{3}{4} = 3\frac{\blacksquare}{\blacksquare} \\
-2\frac{1}{8} = -2\frac{1}{8} \\
\hline
1\frac{\blacksquare}{\blacksquare}
\end{array}
$$

**Find each difference. Simplify.**

4.  $6\frac{1}{3}$ $-2\frac{1}{2}$

5.  $4\frac{2}{5}$ $-1\frac{3}{4}$

6.  $4$ $-1\frac{1}{3}$

7.  $1\frac{1}{8}$ $-\frac{5}{8}$

8.  $6\frac{1}{6}$ $-4\frac{1}{3}$

9.  $4\frac{2}{3}$ $-2\frac{3}{4}$

10. $3\frac{3}{8} - 1\frac{1}{2}$

11. $6 - \frac{3}{4}$

12. $18\frac{1}{4} - 2\frac{5}{6}$

13. $20 - 15\frac{1}{4}$

14. $5\frac{1}{3} - \frac{9}{10}$

15. $4\frac{3}{5} - 2\frac{1}{3}$

16. $5 - 2\frac{1}{2}$

17. $6\frac{7}{8} - 3\frac{3}{8}$

18. $1\frac{1}{2} - \frac{1}{8}$

19. $2\frac{3}{4} - 1\frac{1}{2}$

20. Find the difference of $7\frac{1}{2}$ and $2\frac{1}{16}$.

21. Subtract $2\frac{3}{4}$ from $4\frac{1}{6}$.

22. What is $8\frac{1}{2} - 2\frac{1}{3}$?

23. What is $3\frac{7}{8} - 1\frac{2}{3}$?

24. **Reasoning** When do you need to rename from a whole number when subtracting mixed numbers?

**FOCUS ON SKILLS**                              Name _____
**Mixed Number Operations**

Show all work for the addition and subtraction problems below, including any regrouping that might be necessary.

1.    $1\frac{2}{5} + 7\frac{3}{4} =$

2.    $2\frac{3}{7} + 3\frac{5}{6} =$

3.    $3\frac{1}{2} + 5\frac{2}{3} =$

4.    $5\frac{1}{5} - 2\frac{3}{5} =$

5.    $6\frac{2}{3} - 1\frac{8}{9} =$

6.    $9\frac{1}{9} - 3\frac{1}{3} =$

For the multiplication and division problems, be sure to show all your work, including changing your answer back into a mixed number.

7.    $2\frac{1}{7} \times 3\frac{2}{5} =$

8.    $4\frac{1}{9} \times 5\frac{1}{5} =$

9.    $2\frac{1}{3} \times 9\frac{2}{3} =$

10.    $6\frac{1}{3} \div 2\frac{5}{7} =$

11.    $17\frac{1}{2} \div 1\frac{3}{4} =$

12.    $9\frac{1}{5} \div 3\frac{1}{5} =$

**FOCUS ON CONCEPTS**                    Name _____
**Mixed Number Operations**

1.    The solution to the problem $9\frac{1}{5} \div 3\frac{1}{5}$ is not 3.  Explain why.

2.    If a student gave an answer of 3 to the problem above, how would you try to explain the correct process.

3.    The following solution is incorrect.  Explain where the error was made and then work the problem correctly.

$$6\frac{1}{6} - 2\frac{2}{5} = 6\frac{5}{30} - 2\frac{12}{30} - 5\frac{15}{30} - 2\frac{12}{30} = 3\frac{3}{30} = 3\frac{1}{10}$$

4.    If the above problem was done by a student, what can you say to that student to correct their error?

5.    The following solution is incorrect.  Explain where the error was made and then work the problem correctly.

$$2\frac{1}{5} \times 5\frac{3}{4} = 10\frac{3}{20}$$

6.    The problems in 1, 3, and 5, are very common.  Why do you think these are common ways to work a problem incorrectly?

# DECIMALS

The idea of decimals connects to so many different areas in mathematics which may be one reason they appear so often in textbooks at all levels. Place value and fractions are two of the most obvious connections with decimals, while money and percents are two other areas of application.

Perhaps it is the many places that you see decimals that cause some of the confusion. When should I change a decimal into a percent or fraction for the problem? Does the instructor want my answer as a decimal or a fraction? Can I use my calculator or do I need to know all the details of the decimals? These are just a few of the questions we are likely to hear when teaching the topic.

Calculators are a great tool and we allow them in our classrooms. However, there are certain fundamental concepts that all future teachers must possess. Understanding decimals is one of those areas. Our students must demonstrate their ability to work with decimals (as well as fractions, percents, and whole numbers) without a calculator before being permitted to use it on an exam. This is done by working with problems like you will see on the following pages.

To best understand decimals, it is helpful to initially focus on the place values in the number and to write the decimals as fractions. These two ideas are explored along with ordering decimals in the *Focus on Skills* sheet. In the *Focus on Concepts* problems, you will again do some writing to explain your thought processes to better understand the foundations of decimals. While working with fractions and decimals on these pages you will be doing so in relation to better understanding decimals. There is another section in this book that deals with fractions, decimals, and percents and the relationships between those three topics.

The text pages are from the Grade 3 book and show an introduction to the topic. Decimals appear in books from grades 3 – 6 and focus on the concept of decimals, operations, rounding and estimation, applications, and relating decimals to fractions and percents. The topic is covered at many levels and in many ways and is a critical concept in K-8 mathematics. Be prepared to spend a lot of your teaching time working with decimals.

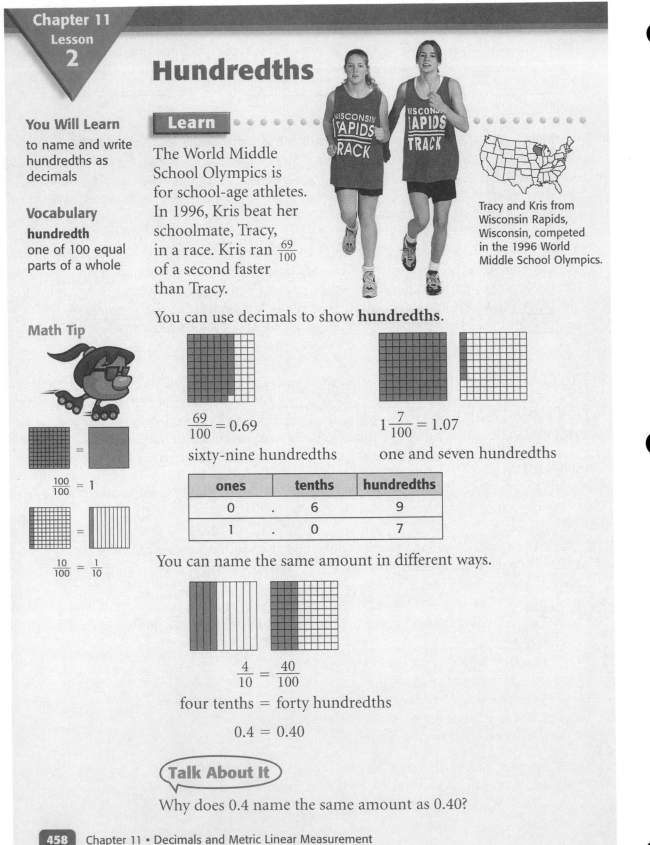

**Chapter 11**
**Lesson**
**2**

# Hundredths

**You Will Learn**

to name and write hundredths as decimals

**Vocabulary**

**hundredth**
one of 100 equal parts of a whole

**Math Tip**

$\frac{100}{100} = 1$

$\frac{10}{100} = \frac{1}{10}$

**Learn**

The World Middle School Olympics is for school-age athletes. In 1996, Kris beat her schoolmate, Tracy, in a race. Kris ran $\frac{69}{100}$ of a second faster than Tracy.

Tracy and Kris from Wisconsin Rapids, Wisconsin, competed in the 1996 World Middle School Olympics.

You can use decimals to show **hundredths**.

$\frac{69}{100} = 0.69$

sixty-nine hundredths

$1\frac{7}{100} = 1.07$

one and seven hundredths

| ones | tenths | hundredths |
|------|--------|------------|
| 0 .  | 6      | 9          |
| 1 .  | 0      | 7          |

You can name the same amount in different ways.

$\frac{4}{10} = \frac{40}{100}$

four tenths = forty hundredths

0.4 = 0.40

**Talk About It**

Why does 0.4 name the same amount as 0.40?

**458**  Chapter 11 • Decimals and Metric Linear Measurement

**Check**

Write the fraction and the decimal to name each shaded part.

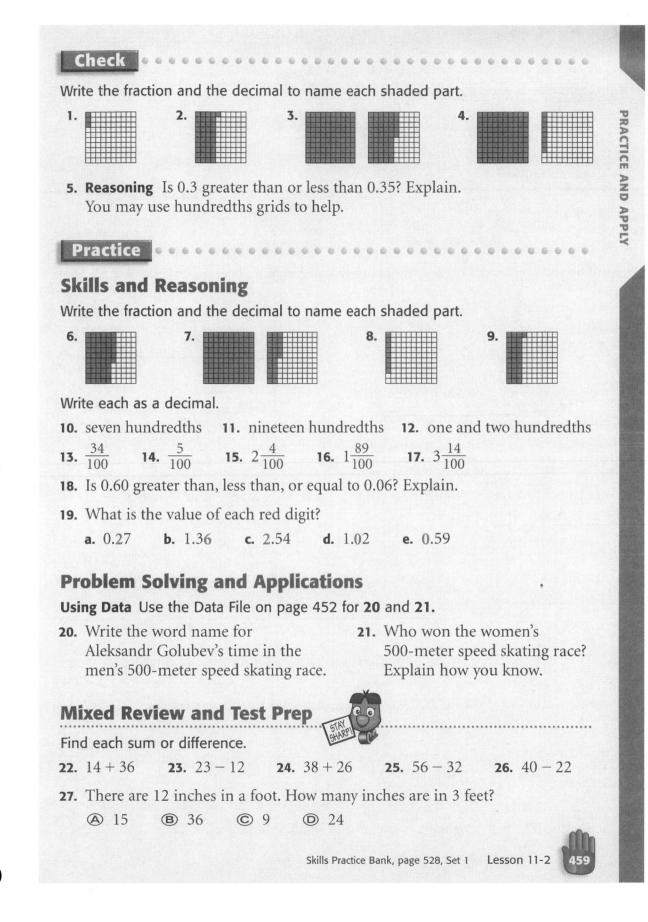

1.

2.

3.

4.

5. **Reasoning** Is 0.3 greater than or less than 0.35? Explain. You may use hundredths grids to help.

**Practice**

## Skills and Reasoning

Write the fraction and the decimal to name each shaded part.

6.

7.

8.

9.

Write each as a decimal.

10. seven hundredths    11. nineteen hundredths    12. one and two hundredths

13. $\frac{34}{100}$    14. $\frac{5}{100}$    15. $2\frac{4}{100}$    16. $1\frac{89}{100}$    17. $3\frac{14}{100}$

18. Is 0.60 greater than, less than, or equal to 0.06? Explain.

19. What is the value of each red digit?

    **a.** 0.27    **b.** 1.36    **c.** 2.54    **d.** 1.02    **e.** 0.59

## Problem Solving and Applications

**Using Data** Use the Data File on page 452 for **20** and **21**.

20. Write the word name for Aleksandr Golubev's time in the men's 500-meter speed skating race.

21. Who won the women's 500-meter speed skating race? Explain how you know.

## Mixed Review and Test Prep

Find each sum or difference.

22. $14 + 36$    23. $23 - 12$    24. $38 + 26$    25. $56 - 32$    26. $40 - 22$

27. There are 12 inches in a foot. How many inches are in 3 feet?

    Ⓐ 15    Ⓑ 36    Ⓒ 9    Ⓓ 24

**FOCUS ON SKILLS**                          Name _____
**Decimals**

For problems 1 – 6, identify which place value to 7 is in.

1.    47.21                                  2.    72.33


3.    10.715                                 4.    321.072


5.    97.1                                   6.    .807


For problems 7 – 10, write each decimal in expanded notation.

7.    48.37                                  8.    981.573




9.    1038.458                              10.    17,083.149




For problems 11- 16, change each decimal to a fraction.  Simplify your answer if necessary.  Show all your work.

11.    .9                                    12.    .87




13.    .383                                  14.    .64




15.    .5                                    16.    .28




17.    Put the following decimals in order from smallest to largest.

       .01     .101    .1001   .1       .110    .001    1.01

**FOCUS ON CONCEPTS**          Name _____
**Decimals**

Changing fractions to decimals involves using division.  For example:

$$\frac{1}{2} = 2\overline{)\,1.0}^{\,.5}$$
$$\underline{-10}$$
$$\quad\; 0$$

and

$$\frac{5}{6} = 6\overline{)\,5.000}^{\,.833}$$
$$\underline{-48}$$
$$\quad 20$$
$$\underline{-18}$$
$$\quad 20$$
$$\underline{-18}$$
$$\quad\; 2$$

We refer to $\frac{1}{2} = .5$ as a terminating decimal since the remainder in the division problem is 0.  We say that $\frac{5}{6} = .833... = .8\overline{3}$ is a repeating decimal.  We note that the 3 continues to repeat by placing 3 dots after the 3 or by putting a bar over the 3.  Repeating decimals occur when you keep getting the same remainder (or sequence of remainders) over and over again.

1.    Show that $\frac{4}{11}$ is a repeating decimal.  Explain why it is a repeating decimal.

For problems 2 – 9, change each fraction to a decimal.  Show your work on a separate sheet of paper.

2.    $\frac{3}{10} =$          3.    $\frac{5}{12} =$          4.    $\frac{9}{11} =$

5.    $\frac{3}{8} =$          6.    $\frac{23}{25} =$          7.    $\frac{4}{7} =$

8.    $\frac{6}{13} =$          9.    $\frac{17}{50} =$

10.    List the fractions that are          11.    List the fractions that are
       terminating decimals.                     repeating decimals.

12.    Describe a process you can use to determine if a fraction converts to a repeating
       or terminating decimal BEFORE you divide.

## OPERATIONS WITH DECIMALS

Hopefully the previous section gave you a better foundation for working with decimals. You should now be ready to perform operations with decimals. And you should be able to do this without your calculator. In this section, the calculator should only be used as confirmation of your answer.

Operations with decimals are not necessarily difficult topics, especially if a solid foundation is present. As with so many other topics in mathematics, when simply presented with an algorithm without any explanation or justification, problems can occur.

You will be looking at the underlying reasons behind why the standard algorithms work as well as asked to explain the reasoning in the *Focus on Concepts* page. Even though this is presented after the *Focus on Skills* sheet, you may want to work on the *Concepts* page first and use those ideas in case you become stuck on the *Skills* sheet. The *Skills* sheet also includes some word problems.

The *Concepts* sheet relies on making a connection between fractions and decimals as well as operations with fractions. Therefore, it is crucial that you, as well as your future students, have a firm understanding of operations with fractions before moving on to decimals. Many students are lost in mathematics in grades 3 – 5 because of the building block relationship between so many of the topics. In order to avoid losing these students you must be prepared to build the solid foundation for those blocks. This can only be done by obtaining a solid foundation for yourself.

Addition and subtraction with decimals occur in the texts as early as Grade 1 in dealing with money. While the book may or may not make a specific reference to decimals when working with money, it is often a student's first introduction to the topic. The text pages shown are from the Grade 4 book and demonstrate addition and subtraction with decimals.

Notice that instead of a totally algorithmic approach, the text pages show a model using manipulatives in order lay a foundation for students. This approach is a common one not only in texts, but in the activities in this manual, and in many classrooms. You should make sure you understand decimals from many different perspectives (physical models, related to fractions, etc.) in order to be an effective teacher.

**Chapter 11**
**Lesson**
**8**

# Exploring Adding and Subtracting Decimals

**Problem Solving Connection**
Draw a Picture

**Materials**
- hundredths grid
- 2 colors of crayons or colored pencils

**Remember**
The decimal 0.2 means 2 tenths, or 20 hundredths.

The decimal 0.02 means 2 hundredths.

**Explore** • • • • • •

You can use grids to show how to add or subtract decimals.

## Work Together

1. Use a hundredths grid to show how to add 0.5 and 0.37.

   a. Shade 5 columns of 10 squares each to show 0.5.

   b. Use a different color. Shade 37 more squares to show 0.37.

   c. Count the shaded squares. How many whole columns are shaded in all? How many extra squares are shaded?

   d. How many hundredths are shaded in all?

   e. Write the decimal for the total squares shaded.

2. Show how to subtract 0.20 from 0.68.

   a. Shade 6 columns and 8 extra squares to show 0.68.

   b. Cross out 2 columns of shaded squares to show 0.20.

   c. Count the squares that are shaded but not crossed out. How many whole columns? How many extra squares?

   d. Write the decimal for total squares shaded but not crossed out.

**Talk About It**

3. What decimal shows the number of squares in a column of the grid?

4. Explain how you can count the grid squares without having to count each one individually.

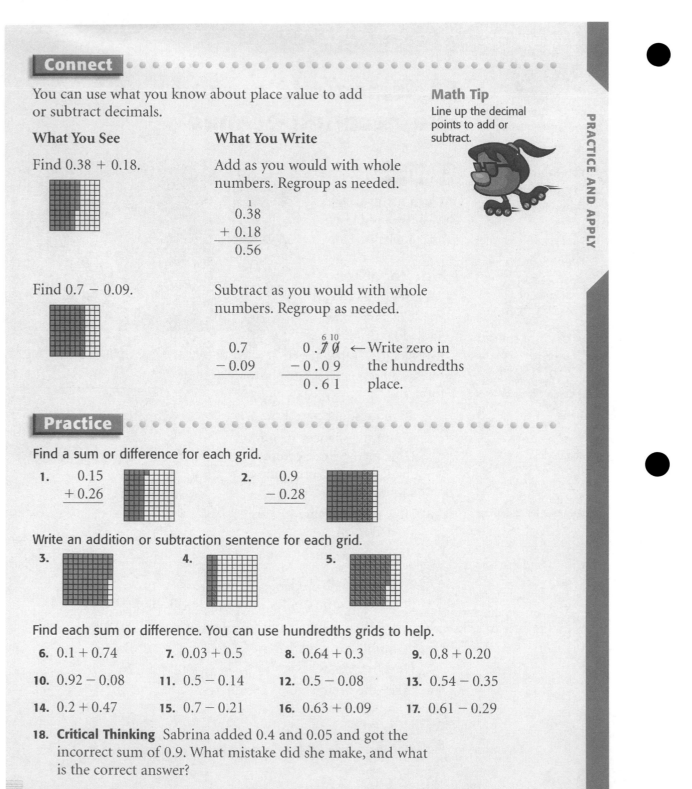

**Connect** • • • • • • • • • • • • • • • • • • • • • • • • • • • • • •

You can use what you know about place value to add or subtract decimals.

**Math Tip**
Line up the decimal points to add or subtract.

**What You See**

Find 0.38 + 0.18.

**What You Write**

Add as you would with whole numbers. Regroup as needed.

$$\begin{array}{r} \overset{1}{0.38} \\ + 0.18 \\ \hline 0.56 \end{array}$$

Find 0.7 − 0.09.

Subtract as you would with whole numbers. Regroup as needed.

$$\begin{array}{r} 0.7 \\ - 0.09 \\ \end{array} \qquad \begin{array}{r} \overset{6\ 10}{0.7\,\cancel{0}} \\ - 0.09 \\ \hline 0.61 \end{array}$$ ← Write zero in the hundredths place.

**Practice** • • • • • • • • • • • • • • • • • • • • • • • • • • • •

Find a sum or difference for each grid.

**1.** $\begin{array}{r} 0.15 \\ + 0.26 \\ \hline \end{array}$

**2.** $\begin{array}{r} 0.9 \\ - 0.28 \\ \hline \end{array}$

Write an addition or subtraction sentence for each grid.

**3.**

**4.**

**5.**

Find each sum or difference. You can use hundredths grids to help.

**6.** $0.1 + 0.74$      **7.** $0.03 + 0.5$      **8.** $0.64 + 0.3$      **9.** $0.8 + 0.20$

**10.** $0.92 - 0.08$      **11.** $0.5 - 0.14$      **12.** $0.5 - 0.08$      **13.** $0.54 - 0.35$

**14.** $0.2 + 0.47$      **15.** $0.7 - 0.21$      **16.** $0.63 + 0.09$      **17.** $0.61 - 0.29$

**18. Critical Thinking** Sabrina added 0.4 and 0.05 and got the incorrect sum of 0.9. What mistake did she make, and what is the correct answer?

**19. Journal** Explain how you can add 0.3 and 0.4 on a hundredths grid.

PRACTICE AND APPLY

**FOCUS ON SKILLS**
**Operations with Decimals**

Name _____

Perform each operation. Show all your work. Do not use a calculator to solve these problems. If your instructor allows, you may check your answers with a calculator when you are finished.

1.     $48.27 + 98.2$                    2.     $33.551 + 170.31$

3.     $81.02 + 18$                       4.     $99.08 - 51.7$

5.     $108.33 - 75.215$                 6.     $378.52 - 191.1$

7.     $.1 \times .31$                     8.     $31.71 \times 5.3$

9.     $19.33 \times 12.2$                10.    $15.5 \div 3.1$

11.    $40.66 \div 2.14$                  12.    $358.36 \div 5.78$

**FOCUS ON CONCEPTS**          **Name** _____
**Operations with Decimals**

Consider the problems .3 + .82 as a fraction problem.  Written as fractions, the problems looks like $\frac{3}{10} + \frac{82}{100}$.  To perform this addition, you need a common denominator of 100.  The problem now looks like $\frac{30}{100} + \frac{822}{100}$.  The answer to this is $\frac{85}{100}$ which is .85 written as a decimal.  This is the same answer you would get just by working with the given decimals.  A similar approach is used for multiplication.  The problem .2 × .51 can be changed to $\frac{2}{10} \times \frac{51}{100}$ which gives you $\frac{102}{1000}$.  Written as a decimal, the answer is .102.

Perform each operation in problems 1 – 8 by changing the decimals to fractions first.  Show all your work.

1.     .5 + .72                2.     .21 + .315                3.     .98 - .7

4.     .273 - .1               5.     .8 × .7                   6.     .11 × .22

7.     .9 ÷ .3                               8.     .25 ÷ .5

9.     Explain the algorithms for addition and subtraction of decimals as it relates to your work in problems 1 – 4.

10.    Explain the algorithm for multiplication of decimals as it relates to your work in problems 5 and 6.

11.    Explain the algorithm for division of decimals as it relates to your work in problems 7 and 8.

# PERCENTS

As most books, including your MET book and K-8 school texts, will tell you, percent is literally translated into "divided by 100" or something similar. The "per" means to divide and "cent" means 100. For a word that has such a simple translation and meaning, there are still many misconceptions associated with the concept.

Perhaps it is the interconnected nature between fractions, decimals, and percents that pose part of the problem. In this section you will explore the concepts associated with percents as fractions and percents as decimals. The relationship between all three topics will be explored in the next section.

Another difficulty with the concept of percents is the application problems. Most students can find 35% of 295 without too much effort. It is when the problems asks for a variation on this problem that the trouble can begin.

Look again at the problem of finding 35% of 295. Most students are taught that "of" means to multiply. So you multiply (.35)(295) to get 103.25. The same approach can be used for:

18 is 30% of what number?

Here, the "of" still means to multiply. But what are you trying to take 30% of? You don't know. So call it "x". Your equation is now $(.30)(x) = 18$ which gives $x = 60$. If you are asked to find

What percent of 375 is 240?

You are trying to take a certain percentage "of" 375. Since you don't know what that percentage is, call it p. Your equation is $(p)(375) = 240$ and the answer for p turns out to be .64. Simply move the decimal two places to the right and your answer is 64%.

This procedure of teaching "of" as multiplication is very straightforward and will always work. You just have to be careful with the terminology, ordering of the question, what you know, and what you're looking for. But by analyzing each problem carefully, application problems should not pose too much difficulty. The *Focus on Concepts* page asks you to work on these types of application problems and show all your work.

The *Focus on Skills* page is meant to give you practice with the basics of percents. You will also apply much of the knowledge and skills used with fractions and decimals. If you are still having difficulty with those topics, this section will prove challenging to you and you may need to go back to fractions and decimals for additional practice.

The text pages are from the Grade 5 book. The approach shows a physical model to go along with the concept of percent. Notice this is also similar to the representation for decimals.

**Chapter 7**
**Lesson**
**12**

# Understanding Percent

**Learn**

In a recent survey, 100 students were asked which household chore they liked the least. The list shows which jobs the students mentioned.

How can you name the part of those surveyed who liked washing dishes the least?

Chores we Like the Least
Washing dishes  31
Laundry  20
Yard work  19
Cleaning room  17
Babysitting  13

Mia wrote a fraction to describe the part.     $\frac{31}{100}$

Donnell used a **percent**. Percent means per hundred or out of 100. % is a symbol for percent.     31%

José thought of the number of students as 31 out of 100.     0.31

**Talk About It**

Look at the grid shown above. What fraction and percent name the unshaded part?

**Check**

Write the hundredths fraction and the percent shaded in each picture.

1.     2.     3.

4. **Reasoning** Doing laundry was the chore least liked by 20 of the 100 students. Use a grid to show the percent who named laundry.

## Practice · · · · · · · · · · · · · · · · · · · · · · · · · · · · · · · · · ·

### Skills and Reasoning

Write the hundredths fraction and the percent shaded in each picture.

**5.**                          **6.**                          **7.**

Write each as a percent.

**8.** 66 out of 100      **9.** $\frac{50}{100}$      **10.** 7 out of 100      **11.** $\frac{48}{100}$

Write each as a hundredths fraction.

**12.** 16%      **13.** 5%      **14.** 59%      **15.** 70%      **16.** 100%

For each set, decide which does **not** belong.

**17.** Ⓐ  6 out of 100      Ⓑ  60%      Ⓒ  $\frac{6}{100}$      Ⓓ  6%

**18.** Ⓐ  29%      Ⓑ  29 out of 100      Ⓒ  2 out of 9      Ⓓ  $\frac{29}{100}$

**Estimation** Estimate the percent of each figure that is shaded.

**19.**                     **20.**                     **21.**

### Problem Solving and Applications

**22. Collecting Data** Survey your class to find the least popular chore.

**Using Data** Use the Data File on page 299 to answer **23** and **24**.

**23.** Where is the most active water found?

**24.** Write a fraction for the amount of active water found in

    **a.** rivers      **b.** lakes      **c.** soil      **d.** atmosphere      **e.** living things

### Mixed Review and Test Prep

Copy and complete. Write >, <, or =.

**25.** $4\frac{4}{5}$ ● $5\frac{8}{10}$      **26.** $2\frac{2}{3}$ ● $\frac{8}{3}$      **27.** $7\frac{3}{5}$ ● $6\frac{2}{3}$      **28.** $3\frac{5}{8}$ ● $3\frac{5}{6}$

**29.** Which number is standard form for eighty-one hundredths?

    Ⓐ  81      Ⓑ  0.081      Ⓒ  8.100      Ⓓ  0.81

**FOCUS ON SKILLS**                    **Name** _____
**Percents**

For problems 1 – 6, change each percent to a fraction and simplify if necessary.  Show all work.

1.      82% =                    2.      65% =                    3.      1% =

4.      37% =                    5.      235% =                   6.      4% =

For problems 7 – 10, change each percent to a decimal.

7.      15% =                                              8.      .5% =

9.      300% =                                             10.     9% =

For problems 11 – 14, change each decimal to a percent.

11.     .2 =                                               12.     .05 =

13.     4 =                                                14.     .007 =

**FOCUS ON CONCEPTS**          Name _____
**Percents**

Set up the necessary equation to solve each problem.  Show all work.  Make sure your answer is **reasonable.**

1.      What is 15% of 97?              2.      What percent of 160 is 70?

3.      200% of what is 90?            4.      Find 93% of 157.

5.      92 is what percent of 800?     6.      82% of what is 32.8?

7.      What is 210% of 50?            8.      What percent of 250 is 150?

9.      50% of what is 150?            10.     98% of 33 is what?

11.     70 is what percent of 14?

## FRACTIONS/DECIMALS/PERCENTS

Relating these three topics seems clear-cut in some ways, yet is a very confusing relationship in many other ways. When students can perform these conversions from one form to another, they have achieved a high level of understanding, even though the computation skills are not very complicated. But with the ability to go from fractions to decimals to percents, in any order, students are demonstrating their knowledge not of basic skills, but of the relationship between the three topics.

Consider the knowledge and skills involved in relating fractions, decimals, and percents. Changing fractions to decimals entails long division and knowing if the division problem will render a repeating or terminating decimal. Changing fractions to percents requires an intermediate step of converting to a decimal and then to a percent. Conversions between decimals and percents involve understanding place value. As going from fractions to percents meant an intermediate step involving decimals, the same can be true for going from percents to fractions. If your strategy is to change the percent to a decimal, that decimal must be converted to a fraction and reduced to its simplest form, if necessary.

So it is not that the skills are difficult or not known by students, it is when all these skills are put together that troubles can occur. The computations, in one form or another, are seen beginning in the grade 3 book. However it is not until the Grade 5 text that all the relationship between the three topics is put together as one big concept. The text pages shown are from the Grade 6 book.

When we teach some of the more complex topics that appear later in the school curriculum, our students who plan to teach kindergarten, $1^{st}$ or $2^{nd}$ grade want to know why it is important to them since they will not be teaching it. Our answer is simple. Even though you will not be teaching students to convert between decimals, fractions, and percents, you will be teaching them the foundations of fractions, even in kindergarten. Without the basic ideas that you start students with, teachers in grades 3 and higher cannot do their job effectively. So no, you will not teach the relationship between these three topics in the very early grades. But don't you think it's necessary to have more knowledge than just what you are teaching? Will your reading level only be at the $2^{nd}$ grade level? Of course not. So why should your math skills?

The *Focus on Skills* page asks you to fill in a table. You are given one of the three forms and must convert to the other two. In *Focus on Concepts* you will explain answers more in depth.

# Connecting Percents to Fractions and Decimals

**Chapter 10 Lesson 10**

**Problem Solving Connection**
- Look for a Pattern

**Materials**
- 10 × 10 grids
- colored pencils or markers

## Explore

Fractions, decimals, and percents all describe portions of a whole. So every percent can be written as both a fraction and a decimal.

## Work Together
### Modeling a Percent

1. Color in a number of squares equal to the percent.

2. Model these percents:
   **a.** 21%   **b.** 55%   **c.** 4%   **d.** 75%

### Modeling a Decimal

3. Color in one column for each tenth.

4. Color in one square for each hundredth.

5. Model these decimals:
   **a.** 0.66   **b.** 0.75   **c.** 0.02   **d.** 0.49

### Modeling a Fraction

6. Divide the grid into groups of equal size. The number of groups should equal the denominator.

7. Color in as many groups as the numerator.

8. Model these fractions:
   **a.** $\frac{3}{4}$   **b.** $\frac{3}{5}$   **c.** $\frac{7}{10}$   **d.** $\frac{1}{2}$

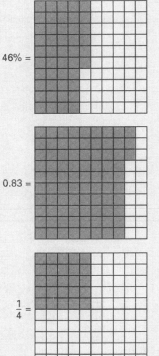

## Talk About It

9. For a given grid, can you describe the number of squares colored using either a percent or a decimal? Explain.

10. For a given grid, can you describe the number of squares colored using either a percent or a fraction? Explain.

## Connect and Learn

Fractions, percents, and decimals all describe parts of a whole. To convert a percent into a fraction or a decimal, rewrite the percent as a fraction over 100.

**Example 1**

**a.** Write 53% as a fraction.

$$53\% = \frac{53}{100}$$

**b.** Write 91% as a decimal.

$$91\% = \frac{91}{100} = 0.91$$

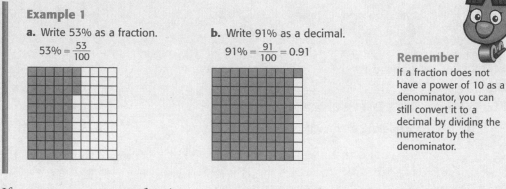

**Remember**

If a fraction does not have a power of 10 as a denominator, you can still convert it to a decimal by dividing the numerator by the denominator.

If you want to convert a fraction to a percent, you can do so with a proportion.

$$\frac{part}{whole} = \frac{percent\ value}{100}$$

**Example 2**

The White's tree frog is $\frac{5}{8}$ the length of the flying gecko. Rewrite this fraction as a percent.

$$\frac{5}{8} = \frac{x}{100}$$

$$8x = 500$$

$$x = 62.5$$

The White's tree frog is 62.5% of the length of the flying gecko.

## Check

Convert each percent to a fraction in lowest terms.

1. 56%
2. 15%
3. 75%
4. 66%
5. 150%
6. 125%

Convert each fraction to a percent.

7. $\frac{5}{8}$
8. $\frac{7}{10}$
9. $\frac{5}{16}$
10. $\frac{4}{8}$
11. $\frac{3}{4}$
12. $\frac{9}{16}$

13. **Reasoning** How can you determine without converting to a percent if a decimal is less than 10%? Greater than 100%?

14. **Reasoning** Which is easier, converting a fraction to a percent or converting a percent to a fraction? Explain.

 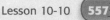

**FOCUS ON SKILLS**
Fractions/Decimals/Percents

Name _____

| Fraction | Decimal | Percent |
|---|---|---|
| $\frac{1}{4}$ | | |
| | .8 | |
| | | 125% |
| | .05 | |
| $\frac{2}{3}$ | | |
| | | $\frac{1}{4}\%$ |

**FOCUS ON CONCEPTS**                    Name _____
**Fractions/Decimals/Percents**

1.    Explain the process you use for changing a percent to a fraction.

2.    Explain why we do not normally go directly from a fraction to a percent.

3.    What do you think is the hardest conversion in the chart from *Focus on Skills*?

4.    Explain your answer to #3.

5.    What do you think is the easiest conversion in the chart from *Focus on Skills* ?

6.    Explain your answer to #5.

7.    Give a situation when a decimal answer is more appropriate than a fraction or a percent answer.

8.    Give a situation when a fraction answer is more appropriate than a percent or a decimal answer.

9.    Give a situation when a percent answer is more appropriate than a decimal or percent answer.

# RATIO AND PROPORTION

Ratio and proportion are so interrelated that they appear together in many texts. However, when they are first introduced in the school texts, they are shown separately to make sure ratios are understood before moving on. For example, the text page shown is from the Grade 5 book and deals just with ratios.

A proportion is simply two ratios that are set equal to each other. In the simplest case of a proportion, it is just two equivalent fractions. However, proportions are most often presented with a piece of the equation missing and it is your job to solve for it. Proportion problems usually occur in the form of a word problem which may be why there are difficulties with the topic. Another problem with proportions is that there can be so many ways to set one up. Look at the following example.

It takes 48 chicken wings to feed Mr. Young's 4$^{th}$ grade class of 20 students. How many wings would be needed for 30 students?

Many students will approach the set up of this problem as it is written and use the ratio $\frac{\text{wings}}{\text{students}}$ which leads to the proportion of $\frac{48}{20} = \frac{x}{30}$. Notice that the number of wings is always in the numerator and number of students is always in the denominator. If you write the ratio you are going to use for solving the problem in words, it will help you set up the necessary proportion.

However, you can also set up a proportion as $\frac{\text{students}}{\text{students}} = \frac{\text{wings}}{\text{wings}}$ or as $\frac{\text{wings}}{\text{wings}} = \frac{\text{students}}{\text{students}}$ or the ratio of $\frac{\text{students}}{\text{wings}}$. You may be asking yourself, "If they all mean the same thing and give me the same answer of 72, what difference does it make?" and that is a good question. In fact, it does not matter which method you choose for solving this problem. What does matter is that you set things up consistently. We have started having our student write the ratio or proportion in words before working with the numbers in order to make sure the correct quantity is in the correct place. So no matter which strategy a student prefers, by writing the words used for solving the problem, they show us what they are doing and are less likely to make a mistake along the way.

*Focus on Skills* is simply setting up ratios and solving proportions. *Focus on Concepts* concentrates on solving proportion word problems.

# Ratios

**You Will Learn**
how to write and
simplify ratios

**Vocabulary**
**ratio**
a pair of numbers
used to compare
quantities

### Learn

A ball game from
New Guinea is played
with 4 players. The
players use 5 balls,
4 hoops, and 1 rope.

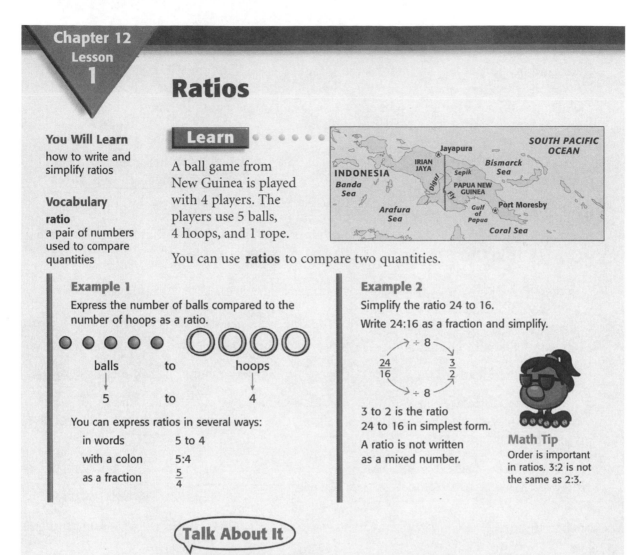

You can use **ratios** to compare two quantities.

**Example 1**

Express the number of balls compared to the
number of hoops as a ratio.

| balls | to | hoops |
|-------|----|-------|
| ↓ | | ↓ |
| 5 | to | 4 |

You can express ratios in several ways:

| in words | 5 to 4 |
|----------|--------|
| with a colon | 5:4 |
| as a fraction | $\frac{5}{4}$ |

**Example 2**

Simplify the ratio 24 to 16.

Write 24:16 as a fraction and simplify.

$$\frac{24}{16} \xrightarrow{\div 8} \frac{3}{2}$$

3 to 2 is the ratio
24 to 16 in simplest form.

A ratio is not written
as a mixed number.

**Math Tip**

Order is important
in ratios. 3:2 is not
the same as 2:3.

### Talk About It

How is a ratio like a fraction? How is it different?

### Check

Use the New Guinea game. Write each ratio in three ways. Simplify.

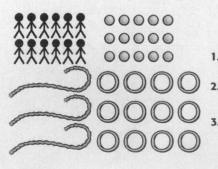

| | | Ratio | | | Ratio Simplified | | |
|---|---|---|---|---|---|---|---|
| **Number of** | ▦ to ▦ | ▦ : ▦ | $\frac{▦}{▦}$ | ▦ to ▦ | ▦ : ▦ | $\frac{▦}{▦}$ |
| 1. | hoops to ropes | 12 to 3 | | | | 4:1 | |
| 2. | ropes to players | | $\frac{3}{12}$ | | | | $\frac{1}{4}$ |

3. **Reasoning** A different game is played with a
   ball-to-player ratio of 1:3. If there are 12 players,
   how many balls will be needed? How do you know?

PRACTICE AND APPLY

## Practice

### Skills and Reasoning

Write each ratio in three ways. Simplify.

**4.** balls to clubs

**5.** tires to tricycle

**6.** hoop to basketballs

**7.** mallets to wickets

**8.** birdies to racquets

**9.** wheels to skates

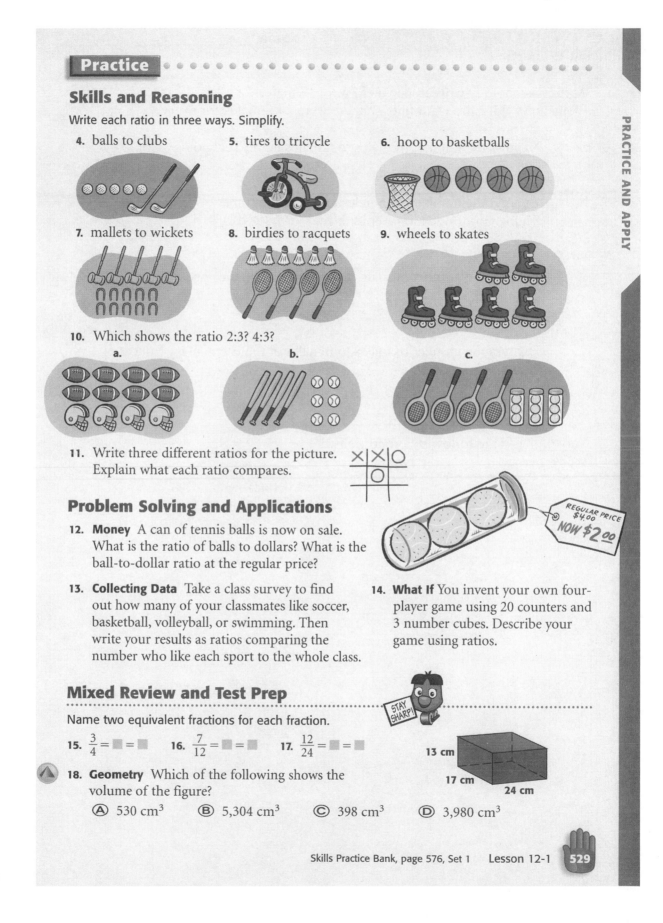

**10.** Which shows the ratio 2:3? 4:3?

a.

b.

c.

**11.** Write three different ratios for the picture. Explain what each ratio compares.

### Problem Solving and Applications

**12. Money** A can of tennis balls is now on sale. What is the ratio of balls to dollars? What is the ball-to-dollar ratio at the regular price?

REGULAR PRICE $4.00   NOW $2.00

**13. Collecting Data** Take a class survey to find out how many of your classmates like soccer, basketball, volleyball, or swimming. Then write your results as ratios comparing the number who like each sport to the whole class.

**14. What If** You invent your own four-player game using 20 counters and 3 number cubes. Describe your game using ratios.

### Mixed Review and Test Prep

Name two equivalent fractions for each fraction.

**15.** $\frac{3}{4} = \blacksquare = \blacksquare$   **16.** $\frac{7}{12} = \blacksquare = \blacksquare$   **17.** $\frac{12}{24} = \blacksquare = \blacksquare$

13 cm   17 cm   24 cm

**18. Geometry** Which of the following shows the volume of the figure?

(A) 530 cm³   (B) 5,304 cm³   (C) 398 cm³   (D) 3,980 cm³

**FOCUS ON SKILLS**                             Name _____
**Ratio and Proportion**

1.    At the Technology Conference there were 17 males for every 12 females.  At the
      Elementary Education Conference there were 24 females for every 14 males.

      a.    Express the number of males to females at the Technology Conference.

      b.    Express the number of females to the total number of people at the
            Elementary Education Conference.

      c.    Express the number of males to the total number of people at the
            Technology Conference.

      d.    Express the number of males to females at the Elementary Education
            Conference.

2.    Solve for $x$ in the following proportions.

      a.    $\dfrac{12}{x} = \dfrac{18}{45}$                    b.    $\dfrac{x}{7} = \dfrac{10}{21}$

      c.    $\dfrac{3}{8} = \dfrac{x}{20}$                      d.    $\dfrac{5}{7} = \dfrac{x}{98}$

**FOCUS ON CONCEPTS**                    Name _____
**Ratio and Proportion**

1.    There are approximately 2 pounds of muscle for every 5 pounds of body weight. How much weight of a 90 pound child is muscle?

2.    If 4 grapefruits sell for 79 cents, how much do 6 grapefruits cost?

3.    David read 40 pages of a book in 50 minutes. How many pages should he be able to read in 80 minutes?

4.    Jim found out that after working for 9 months he had earned 6 days of vacation time. How many days per year does he earn?

5.    Jeannie takes inventory of her closet and discovers that she has 8 shirts for every 5 pair of jeans. If she has 40 shirts, how many pairs of jeans does she have?

6.    The directions on a package of punch mix say to use 4 quarts of water for every 3 tablespoons of punch mix. How many gallons of punch can be made using 39 tablespoons of punch mix?

7.    An eight ounce bag of tater tots says that it will feed 10 people. If there are 35 people coming to a party, how many bags of tater tots are needed?

8.    In the English alphabet, determine the ratio of vowels to consonants.

9.    Write a word that has a ratio of 2 vowels for every 3 consonants

# DESCRIBING DATA

Our everyday lives make up a vast amount of data, for instance, types of clothing we wear, food we eat, and pets we own. In kindergarten, students sort data by shape, size, color, and other characteristics. As they progress through the elementary grades, they describe the data using various numbers such as mean, median, and mode. By Grade 6 they determine the "spread" of the data and begin looking for outliers in the data.

The sample pages are from a Grade 6 book. The activities on these pages have students identify outliers in the data sets. Also, the students are asked to find the mean, median, and mode of the data set with and without the outliers and to describe the effect outliers have on these values.

The *Focus on Skills* problems encourage you to become proficient at calculating the measures of central tendency (mean, median, mode) and the measures of variability (range, quartiles, outliers, standard deviation). Although elementary grade students will not be asked to calculate values such as quartiles and standard deviation, as a future teacher of these students, you should understand the concepts and be able to perform the calculations.

The *Focus on Concepts* questions encourage you to go beyond the calculations and to make sense of what the values tell us about the data. These concepts enable us to interpret data sets and to draw conclusions about a given situation. This important step is sometimes omitted or overlooked in mathematics classrooms. Often, our students are great at performing the calculations, but have little idea of why they perform them or what they tell them about the data.

There are many reasons why you should understand these skills and concepts. Of course an obvious reason is you must have a deeper understanding of the concepts that you teach to your students. A second reason is best given as an example. Suppose you are teaching sixth grade and have several classes take the same test. You want to analyze the test grades (for yourself, the principal, or parents). How well did a student perform compared to the other students in the class? How well did the overall class compare to the other classes that took the exam? These types of questions can be answered by describing the data using the measures of central tendency and variability discussed on this page. Thus, out of concern for your students' education, you must master the ability to describe data.

# The Effects of Outliers

**You Will Learn**

■ to determine if an outlier affects the analysis of a data set

**Vocabulary**

**outlier**

**Learn** • • • • • • • • • • • • • • • • • • • • • • • • • • • • • • • •

Measures such as mean, median, and mode can be affected by data items that are much different from the other items in a set.

An **outlier** is a number in a data set that is very different from the rest of the numbers. Outliers can have a major effect on the mean.

In the last lesson, you saw that the mean of a data set may represent the set well. For example, the mean of the daily high temperatures shown here is 91°F. Because the mean is close to all of the data, it represents the set well.

| Daily High Temperatures (°F) | |
|---|---|
| Monday | 88 |
| Tuesday | 94 |
| Wednesday | 94 |
| Thursday | 92 |
| Friday | 87 |

Suppose that on Saturday the temperature plunges to 55°F. Look what happens to the mean:

$$88 + 94 + 94 + 92 + 87 + 55 = 510$$

$$510 \div 6 = 85$$

The mean temperature of 85°F is *less than* five of the six data items. It has been pulled downward by the outlier, 55°F.

The table shows that the median is affected only slightly by the addition of the Saturday outlier. The mode hasn't changed.

| | Mon–Fri | Mon–Sat |
|---|---|---|
| **Mean** | 91 | 85 |
| **Median** | 92 | 90 |
| **Mode** | 94 | 94 |

**Did You Know?**
The highest temperature ever recorded in the United States was 134°F in Death Valley, California, on July 10, 1913.

You can see that a data set with an outlier is usually better represented by the median or the mode. The mean is often pulled too far toward the outlier to represent the set well.

**Example 1**

Find the median, mode, and mean of the data with and without the outlier.

| Tallest Buildings in Las Vegas | |
|---|---|
| Building | Height (ft) |
| Vegas World Tower | 1,012 |
| Fitzgerald Hotel | 400 |
| Landmark Hotel | 356 |
| Las Vegas Hilton | 345 |

<u>Without outlier</u>

Median: 356

No mode

Mean: 400 + 356 + 345 = 1,101

1,101 ÷ 3 = 367

<u>With outlier</u>

Median: 400 + 356 = 756

756 ÷ 2 = 378

No mode

Mean: 400 + 356 + 345 + 1,012 = 2,113

2,113 ÷ 4 = 528.25

The degree to which an outlier affects the mean is usually significant. Occasionally, it is not.

### Example 2

Find the median, mode, and mean of the data with and without the outlier.

| Normal January Temperatures in Selected U.S. Cities | | | |
|---|---|---|---|
| Galveston, TX | 53°F | San Francisco, CA | 49°F |
| Mobile, AL | 50°F | Savannah, GA | 49°F |
| San Antonio, TX | 49°F | Houston, TX | 50°F |

In this data set, 53°F is the outlier because all of the other temperatures are either 49°F or 50°F.

Without outlier

Median: 49°F

Mode: 49°F

Mean: 50 + 49 + 49 + 49 + 50 = 247

247 ÷ 5 = 49.4°F

With outlier

Median: 49 + 50 = 99

99 ÷ 2 = 49.5°F

Mode: 49°F

Mean: 53 + 50 + 49 + 49 + 49 + 50 = 300

300 ÷ 6 = 50°F

When the outlier was included, the mean temperature increased by less than one degree, the median temperature increased by less than one degree, and the mode was unchanged.

**Talk About It**

1. Why doesn't the mode change when an outlier is added to a data set?

2. Would a high outlier and a low outlier affect a data set differently? If so, how?

**Check** • • • • • • • • • • • • • • • • • • • • • • • • • • • • • • • • • • • • • • • •

1. Find the median, mode, and mean with and without the outlier.

| States with Most Indian Reservations | | | | | | | | |
|---|---|---|---|---|---|---|---|---|
| State | AZ | CA | MN | NV | NM | WA | WI | SD |
| Number | 23 | 96 | 14 | 19 | 25 | 27 | 11 | 9 |

2. **Reasoning** What relationship is shared by an outlier and the range of a set of data?

## Practice

### Skills and Reasoning

Identify the outlier in each data set.

**3.** 24, 24, 18, 56, 25, 12, 15, 22

**4.** 34, 28, 31, 34, 2, 29, 21

**5.** 7, 6, 9, 10, 11, 6, 8, 11, 0, 10, 7, 8

**6.** 200, 225, 3,000, 500, 325, 311

Identify the outlier in each data set.

**7.**

| Stem | Leaf |
|------|------|
| 0 | 3 |
| 1 | 0 0 0 1 1 5 8 |
| 2 | 1 3 3 8 9 |
| 3 | 0 0 |

**8.**

```
                        x
                        x   x
                    x   x   x
    x   x   x   x   x                           x
  +---+---+---+---+---+---+---+---+---+---+--->
    0   1   2   3   4   5   6   7   8   9
```

Use the Temperature chart to answer **9.**

**9.** Find the median, mode, and mean of the temperatures with and without the outlier.

| Normal June Temperatures in Corville | | | | | |
|------|------|------|------|------|------|
| 1993 | 1994 | 1995 | 1996 | 1997 | 1998 |
| 71°F | 73°F | 72°F | 79°F | 73°F | 73°F |

Use the British Open Tournament Scores chart to answer **10** and **11.**

**10. Estimation** Estimate the mean, median, and mode with and without the outlier, then find them by computing. How did your estimates compare to your computations?

**11.** Did the outlier affect the mode? The mean? The median? Which did it affect the most?

| British Open Tournament Scores (1996) | |
|------|------|
| John Daly | 282 |
| Costantino Rocca | 282 |
| Michael Campbell | 283 |
| Steven Bottomley | 283 |
| Barry Lane | 288 |

Use the Dinah Shore Tournament Scores to answer **12.**

**12. a.** Find the mean, median, and mode with and without the outlier.

**b.** Did the outlier affect the mode? The mean? The median? Which did it affect the most?

| Dinah Shore Tournament Scores (1996) | |
|------|------|
| Nanci Bowen | 285 |
| Susie Redman | 286 |
| Brandie Burton | 287 |
| Sherri Turner | 287 |
| Meg Mallon | 292 |

**13. Critical Thinking** Consider the data sets 20, 20, 100 and 450, 450, 460. Without computing, in which set of data will the outlier most affect the mean? Explain.

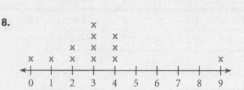

## Problem Solving and Applications

The following table shows the number of games Michael Jordan has played with the Chicago Bulls.

| Year | Games Played | Year | Games Played | Year | Games Played |
|------|------|------|------|------|------|
| 1984–85 | 82 | 1988–89 | 81 | 1992–93 | 78 |
| 1985–86 | ??? | 1989–90 | 82 | 1993–94 | 78 |
| 1986–87 | 82 | 1990–91 | 82 | 1994–95 | 17 |
| 1987–88 | 82 | 1991–92 | 80 | 1995–96 | 82 |

14. Find the mean, median, and mode of the data provided. Ignore the entry for 1985–1986.

15. During the 1985–86 season, Michael had an injury and played only 18 games. Add this outlier to the data and recompute the mean, median, and mode.

16. How do the outliers affect the mean? The median? The mode?

17. **Critical Thinking** Which number (the median, the mode, or the mean) is the best measure to use when describing the number of games Michael played each year? Explain.

18. **Measurement** Measure the feet of your classmates. Organize the data. What are the outliers of the data?

19. **Patterns** Consider the data in these sets. Look for a pattern.
    2, 3, 15          4, 9, 60          8, 27, 240          16, 81, ?
    What is the outlier in the fourth set of data?

## Mixed Review and Test Prep

Perform the appropriate operation.

20. $234 + 5{,}278$

21. $5{,}678 - 3{,}991$

22. $26 \times 52$

23. $306 \div 9$

24. $43{,}675 + 2{,}344$

25. $89{,}021 - 5{,}811$

26. $329 \times 86$

27. $915 \div 30$

28. Does there appear to be a trend in the scatterplot to the right? Explain.

29. Which has the greatest value for the following set of data, the mean, the median, the mode, or the range?
    94, 88, 11, 90, 94, 92
    Ⓐ Mean  Ⓑ Median  Ⓒ Mode  Ⓓ Range

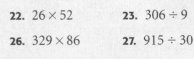

**Hits For Women's Softball Players**

Number of hits / Number of games played

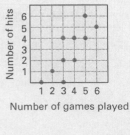

**FOCUS ON SKILLS**                                    **Name**_____
**Describing Data**

The following are test scores for Dr. Lanier's algebra class:

| 91 | 90 | 79 | 83 | 93 | 97 | 72 | 88 | 89 | 65 |
|----|----|----|----|----|----|----|----|----|----|
| 85 | 87 | 70 | 93 | 75 | 75 | 71 | 78 | 89 | 30 |

Calculate or identify the following values for the data set.

1.   mean

2.   median

3.   mode

4.   minimum value

5.   lower quartile

6.   upper quartile

7.   maximum

8.   range

9.   interquartile range

10.  outliers (if any)

11.  standard deviation

**FOCUS ON CONCEPTS**                    Name_____
**Describing Data**

You may use the data set and the values from the Skills sheet to help you answer the following.

Focus on mean, median, and mode:

1.  Which of the three measures (mean, median, or mode) best describes the data set on the *Skills* sheet?  Explain your answer.

2.  Using your own words, explain (a) the meaning of each term, (b) how to calculate or identify each value, and (c) what the value tells us about a data set.

    - Mean

    - Median

    - Mode

3.  Give an example of a set of data for which the mean would best describe the data.  Explain why the mean is the best descriptor.

4.  Give an example of a set of data for which the median would best describe the data.  Explain why the median is the best descriptor.

5.  Give an example of a set of data for which the mode would best describe the data.  Explain why the mode is the best descriptor.


Focus on outliers and measures of variability:

6.  Using your own words, explain (a) the meaning of each term, (b) how to calculate or identify each value, and (c) what the value tells us about a data set.

    - quartiles
    - standard deviation
    - outlier


7.  What effect does an outlier have on the (a) mean, (b) median, and (c) mode of a data set?

# DISPLAYING DATA AND INTERPRETING GRAPHS

We are surrounded in our daily lives by data and graphs (pictures) that represent that data. For instance, some electric companies send out bills that include a bar graph depicting energy usage for the past 13 months. A popular asthma medication has an insert that requires patients to "chart" (draw a line graph of) their response to the drug. Many newspapers, magazines, and websites display data as some type of graph (bar graph, circle graph, line plot, pictograph, etc.) on a daily basis. A number of these graphs attempt to convince us to buy a certain product, invest in a certain stock, or use the services offered by certain group. Other graphs give us a information about our environment such as city crime rate or local school performance. Often, we make decisions based upon the graphs and data we see in these publications. Thus, if we are to make accurate decisions, we must be able to interpret these images and draw appropriate conclusions.

The sample pages are from a Grade 6 book. The activities from the Grade 6 book have students read and interpret information from the graph. They are asked to determine if a graph is misleading and, if so, how they would correct the graph. Also, they are encouraged to think about why someone would create a misleading graph.

For you to correctly interpret graphs, you must have an understanding of how the graphs are constructed. Students are taught how to make pictographs as early as kindergarten. As they progress through the next several years, they learn to construct bar graphs. During the later elementary grades, they construct circle graphs, line graphs, line (dot) plots, stem-and-leaf diagrams, box-and-whisker plots, and scatter plots. Histograms are usually introduced in middle school. Hence, the *Focus on Skills* exercises encourage you to construct various graphs that are taught in elementary and middle school.

The *Focus on Concepts* questions encourage you to practice reading and interpreting various data displays. They also encourage you to look for misleading graphs and information. Throughout the elementary and middle school grades, students are taught to read graphs and make predictions. So it is very important that their teachers are proficient at the same tasks.

Students often ask, "When are we ever going to use this?" The abilities to read and interpret graphs and to make decisions and predictions based on these graphs are some of the essential skills needed in life.

# Misleading Graphs

**You Will Learn**

■ to identify common ways that a graph can suggest misleading relationships

**Did You Know?**

Great white sharks are known to attack humans, but they usually don't eat them. Great white sharks usually prey on seals, sea lions, whales, and other sharks.

**Learn** ● ● ● ● ● ● ● ● ● ● ● ● ● ● ● ● ● ● ● ● ● ● ● ● ● ● ●

There are many ways to make a graph that can mislead a reader. One way is to start labeling the graph at a number other than zero without indicating that some numbers have been skipped.

**Example 1**

Is the great white shark twice as long as the mako shark?

In graph A, the top bar is twice as long as the bottom bar. But the value for the great white shark, 16, is not twice the value for the mako shark, 13.

In graph B, the great white shark is clearly not twice as long. When the bar graph starts at 0, the graph is not misleading.

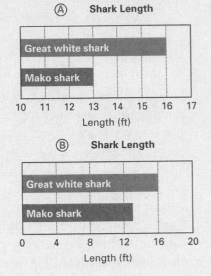

Sometimes the values shown in the vertical scale may not form a consistent pattern.

**Example 2**

Is the hippo able to hold its breath for twice as long as the sea otter?

Both bars start at zero, and the hippo bar is twice as tall as the sea otter bar. But the data values show that a sea otter can hold its breath for 5 minutes and the hippo for 15 minutes—three times as long as a sea otter. A misleading impression is created because the 5–15 space covers more values than the 0–5 space, but both spaces have equal heights.

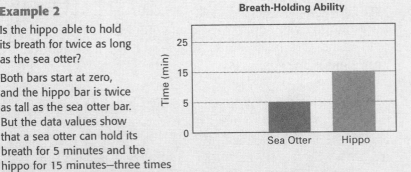

A graph can also lengthen or shorten the spaces between data values in order to mislead readers.

### Example 3

Which admission price went up more quickly?

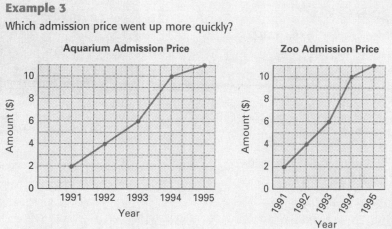

**Math Tip**
Many graphs can be misleading because the vertical scale has been drawn incorrectly. When evaluating a graph for misleading impressions, check the vertical scale first.

In the graph on the right, the years are much closer together, so the line appears to climb more rapidly. However, both graphs show exactly the same data. Neither admission price went up more quickly.

### Talk About It

1. How can a graph's labels be manipulated to mislead the reader?

2. How can the axes of a graph be manipulated to mislead the reader?

### Check

Tell how each graph could create a misleading impression.

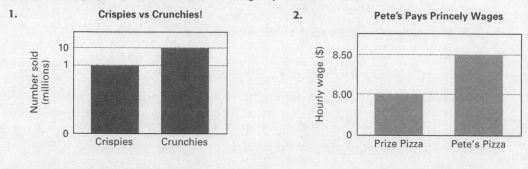

1. **Crispies vs Crunchies!**

2. **Pete's Pays Princely Wages**

3. **Reasoning** Why might someone want to create a misleading graph? Give examples from everyday life.

**Practice**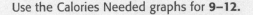

## Skills and Reasoning

Use the Life Span graph for **4–8.**

4. What is the information in the graph about?

5. **Estimation** How many times greater does the manatee's life span appear to be than the dolphin's?

6. Read the graph. What is the approximate life span of the dolphin? The manatee?

7. What is the difference in the life spans of the manatee and the dolphin?

8. **Critical Thinking** Is the bar graph misleading? If so, how would you correct the graph?

Florida is the home for most manatees in the U.S. In 2000, only 2,222 were counted off the coasts of Florida

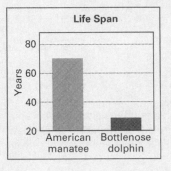

Use the Calories Needed graphs for **9–12.**

9. **Science** How many calories a day does a mouse need? A robin?

10. **Estimation** For each graph, the robin's calorie bar appears to be how many times greater than the mouse's calorie bar?

11. **Critical Thinking** Do you think either graph is misleading? Explain.

12. How many calories does each animal need in a 30-day month?

13. **Using Data** Use the Lengths of Sharks graph on page 2 to answer each question.

   a. What is the difference in length between the basking shark and the mako shark?

   b. Which shark is one-half the length of the basking shark?

PRACTICE AND APPLY

## Problem Solving and Applications

Use the population graphs to answer 14–17.

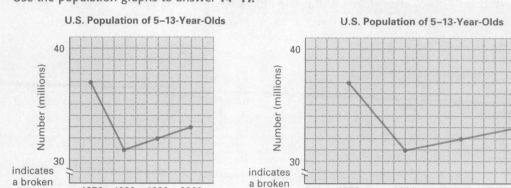

14. **Critical Thinking** How many more 5–13-year-olds will there be in the year 2000 than there were in the year when their population was the least?

15. **Patterns** What do you think the population of 5–13-year-olds will be in the year 2010?

16. **Reasoning** Why might someone want to represent the information with the second graph?

17. **Write Your Own Problem** Use the population data to write a problem. After solving your problem, exchange it for a classmate's. Then compare answers.

## Mixed Review and Test Prep

Write each number in words.

18. 639        19. 204        20. 883        21. 913

22. 6,728      23. 8,912      24. 2,856      25. 1,045

Subtract.

26. 239 − 51   27. 681 − 67   28. 714 − 80   29. 809 − 37

30. 489 − 211  31. 503 − 432  32. 932 − 601  33. 883 − 577

34. How many times must you rename when you subtract 325 from 4,037?
    Ⓐ Zero times    Ⓑ One time    Ⓒ Three times    Ⓓ Four times

**FOCUS ON SKILLS**                                    Name_____
**Displaying Data and Interpreting Graphs**

1. Joyce has a habit of dropping change into the bottom of her purse. While cleaning out her purse, she discovered 48 pennies, 12 nickels, 26 dimes, and 14 quarters.

   Display the contents of Joyce's purse as a

   - Dot plot

   - Pictograph

   - Bar graph

   - Circle graph

   *Note: Be sure to include titles and labels on your graphs.

2. The following table shows Susie's kilowatt hour (KWH) usage each month for the year 2002.

   **Susie's Energy Usage for the 2002 Year**

   | Month | KWH (hundreds) | Month | KWH (hundreds) | Month | KWH (hundreds) |
   |---|---|---|---|---|---|
   | January | 1014 | May | 710 | September | 1150 |
   | February | 1220 | June | 990 | October | 975 |
   | March | 820 | July | 1000 | November | 407 |
   | April | 415 | August | 1100 | December | 1000 |

   Display Susie's energy usage for the year as a line graph.

**FOCUS ON CONCEPTS**                     Name_____
**Displaying Data and Interpreting Graphs**

The following graph displays the monthly credit card spending for the Smith household for the 2002 year.

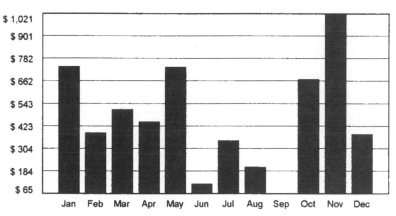

**Monthly Spending Summary**

Use the graph to answer the following questions.

1.    Approximately how much was spent in the month of April?  September? December?

2.    In which month was the most amount spent?  least amount?

3.    What is the approximate difference in the amounts spent in June and July?

4.    Write a paragraph describing the Smith's credit card spending for the year.

5.    What type of graph is presented?

6.    Does the graph have any missing components?  Explain.

7.    Is the graph misleading in any way?  Explain.

8.    Refer to the second problem (Susie's energy usage) on the *Skills* sheet. Reflect upon your line graph and write a paragraph describing Susie's energy usage for the year.

9.    Why do you think she only used 415 KWH in April and 407 KWH in November?  Write several conjectures.

## PROBABILITY

What are the chances it will rain today?  What are the chances I will win the lottery?  What are the chances I will pass my math test?  Our conversations are filled with the phrase "what are the chances."  When asking these questions we are really implying "what is the probability" that something is going to happen.  Writing simple probabilities is introduced as early as Grade 1.  Other probability ideas are scattered throughout the elementary and middle school curriculum.  In the early grades students learn about certain, possible, and impossible events, fair and unfair games, and outcomes.  By Grades 4 and 5 they are drawing tree diagrams, conducting simulations, and calculating experimental as well as theoretical probabilities.  In middle school they study counting principles and distinguish between independent and dependent events.  Throughout the curriculum they are challenged to make predictions based upon their findings.

The sample pages are from a Grade 2 book.  The activities on these pages ask students to determine which color is more likely to be chosen from a bag containing various amounts of different colored cubes.  They determine this more-likely color by selecting a cube from the bag, recording its color, placing the cube back into the bag, and repeating the entire process for a given number of times.  After exploring with "pre-made" bags, students are asked to "create" bags so that a particular color is more likely to be chosen.  It is important to note here that the students are conducting an "experiment" and recording the "outcome" of each "trial."   Although the formal vocabulary is not used and the students are not asked to "write" a probability, the activities on these pages address some very important concepts in probability.

The *Focus on Skills* exercises encourage you to write probabilities for a specific single-stage experiment.  Some common single-stage experiments are tossing a coin, rolling a die, and drawing a card out of a deck.

The *Focus on Concepts* exercises are in two parts.  The first part encourages you to find probabilities and answer questions about a multi-stage experiment.  Often, students have difficulty finding the probabilities for these experiments.  Multi-stage experiments, such as drawing three coins out of a bag without replacement, involve tree diagrams and conditional probabilities.  They provide an opportunity for a discussion of the concepts of independent and dependent events.  Also, these types of problems are great examples for illustrating to students the necessity for reading the problem carefully.  Overlooking the words "with replacement" or "without replacement" is a fatal error in answering the questions correctly.

The second part of the *Concepts* activities requires you to perform a simulation of shooting a basketball. You, as well as your future students, need practice in developing models of real-life situations. Some models that people create adequately represent a given situation while others do not.  You are asked to simulate the activity and then reflect upon the accuracy of the model to represent the activity.

Name _____

**Explore Probability**

**Explore** ● ● ● ● ● ● ● ● ● ● ● ● ● ● ● ● ● ● ● ● ● ● ● ● ● ●

Which color are you more likely to pick?

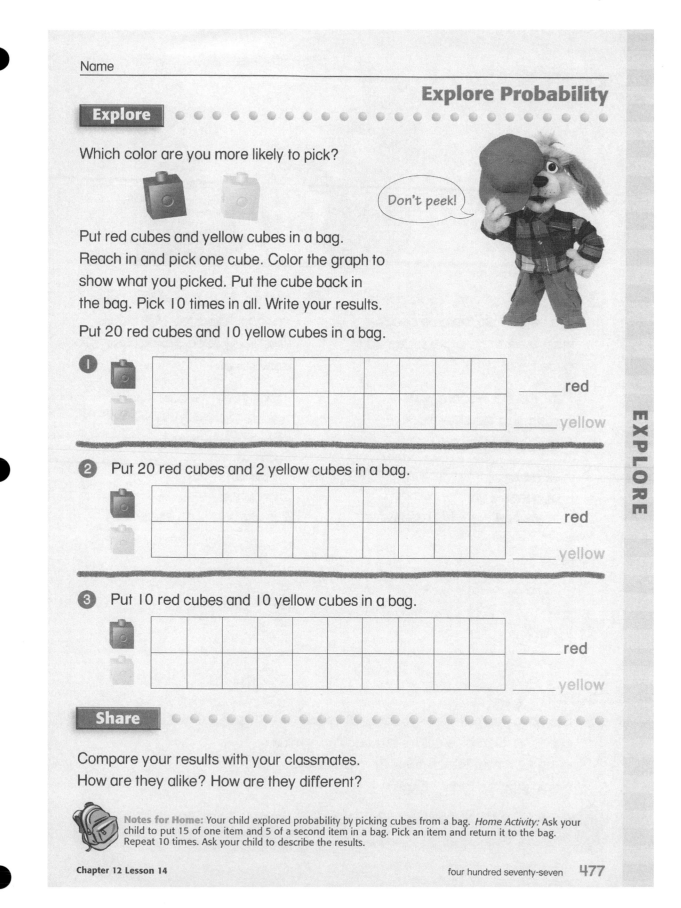

Put red cubes and yellow cubes in a bag.
Reach in and pick one cube. Color the graph to
show what you picked. Put the cube back in
the bag. Pick 10 times in all. Write your results.

*Don't peek!*

Put 20 red cubes and 10 yellow cubes in a bag.

1
_____ red
_____ yellow

2  Put 20 red cubes and 2 yellow cubes in a bag.
_____ red
_____ yellow

3  Put 10 red cubes and 10 yellow cubes in a bag.
_____ red
_____ yellow

**EXPLORE**

**Share** ● ● ● ● ● ● ● ● ● ● ● ● ● ● ● ● ● ● ● ● ● ● ● ● ● ●

Compare your results with your classmates.
How are they alike? How are they different?

**Notes for Home:** Your child explored probability by picking cubes from a bag. *Home Activity:* Ask your
child to put 15 of one item and 5 of a second item in a bag. Pick an item and return it to the bag.
Repeat 10 times. Ask your child to describe the results.

**Connect**

There are fewer yellow cubes in this bag. It's least likely you'll pick yellow!

**EXPLORE**

④ Put red, blue, and yellow cubes in a bag. Make the blue cubes least likely to be picked. Use 20 cubes in all.

Pick 1 cube. Record your results. Put the cube back. Pick 20 times in all.

| Number of cubes in my bag | My results |
|---|---|
| R _____ | R _____ |
| B _____ | B _____ |
| Y _____ | Y _____ |

⑤ Put red, blue, and yellow cubes in a bag. Make the blue cubes most likely to be picked. Use 20 cubes in all.

Pick 1 cube. Record your results. Put the cube back. Pick 20 times in all.

| Number of cubes in my bag | My results |
|---|---|
| R _____ | R _____ |
| B _____ | B _____ |
| Y _____ | Y _____ |

**Journal**

⑥ Suppose you put 10 red cubes, 10 yellow cubes, and 10 blue cubes in a bag. You pick a cube 12 times. Which color do you think you would pick the most? Explain.

**Notes for Home:** Your child put red, blue, and yellow cubes in a bag and predicted which color was most likely or least likely to be picked. *Home Activity:* Ask your child to put 3 red crayons and 1 blue crayon in a bag. Have your child predict which color is more likely to be picked.

**FOCUS ON SKILLS**                              Name_____
**Probability**

Joyce's purse contains 48 pennies, 12 nickels, 26 dimes, and 14 quarters.   She draws one coin from the purse.

1.    Which coin is most likely to be chosen?

2.    Which coin is least likely to be chosen?

3.    What is the probability she will get a penny?

4.    What is the probability she will get a nickel?

5.    What is the probability she will get a dime?

6.    What is the probability she will get a quarter?

7.    What is the probability she will get a fifty cent piece?

8.    What is the probability she will get a penny and a dime?

9.    What is the probability she will get a nickel or a quarter?

10.    What is the probability she will get a penny, nickel, dime, or quarter?

**FOCUS ON CONCEPTS**                                    **Name**_____
**Probability – Multistage Events**

Joyce removes the pennies from her purse.  Now her bag contains 12 nickels, 26 dimes, and 14 quarters.   She draws three coins from the purse (one at a time without replacement).

Draw a tree diagram that represents Joyce's actions and label the probabilities on the branches.

1.    What is the probability that Joyce gets a dime on the first draw?

2.    What is the probability that Joyce gets a dime on the second draw, given she got a dime on the first draw?

3.    What is the probability that Joyce gets a dime on the third draw, given she got a nickel on the first draw and a quarter on the second draw?

4.    What is the probability that Joyce gets a dime on the first draw, a nickel on the second draw, and a quarter on the third draw?

5.    What is the probability that Joyce gets a dime, nickel, and quarter?

6.    What is the probability that Joyce gets all quarters?

7.    What is the probability that Joyce gets all pennies?

8.    What is the probability that Joyce gets three of the same type coin?

9.    What is the probability that Joyce gets more than fifteen cents?

10.    What is the probability that Joyce gets enough money to buy a can of diet coke that costs sixty cents?

**FOCUS ON CONCEPTS**                                    **Name**_____
**Probability - Simulations**

Simulation:   Basketball Free Throws

You will simulate shooting basketball free throws.  Using any coin, let heads represent the ball going into the basket and tales represent the ball missing the basket. Each toss of the coin represents a shot at the free throw line.  Simulate shooting 35 free throws.  Make a chart on your paper similar to the one below to record your results.  For each trial (toss), place an x in the appropriate column.

| Toss number | Went in (Heads) | Missed (Tails) |
|:-----------:|:---------------:|:--------------:|
| 1           | x               |                |
| 2           |                 | x              |
|             |                 |                |

Questions:

1.  How many free throws went into the basket?

2.  How many free throws missed the basket?

3.  What is the experimental probability that you will make the next free throw?

4.  What is the experimental probability that you will miss the next free throw?

5.  This simulation assumes that you have an equal chance of making the free throw or missing the free throw.  Is this a correct assumption?  Why?

6.  How would you change this simulation to better predict a person's abilities at making free throws?

## LINES AND ANGLES

We can describe many of the things in our world using lines and angles. A pencil can be described as a segment. The stripes on a highway can be described as parallel lines. The angle formed by the edge of one wall and the edge of the floor can be described as a right angle. Two or more intersecting roads can be described as intersecting lines, and various types of angles can be identified (vertical, adjacent). In addition, some of these intersecting roads can be described as parallel or perpendicular. The iron bars on a ferris wheel coming out from the center can be thought of as forming acute angles. We can go on and on with our list.

Lines, segments, rays, and angles are found throughout the elementary and middle school curriculum. They are given formal definitions by third grade. As students progress through the grades these terms are reinforced and connected to other topics in the curriculum. For instance, students learn to recognize shapes such as triangles in kindergarten. They identify segments and different types of angles in third grade. In subsequent grades these concepts are combined and students discuss the angles and the sides (segments) of a triangle.

The sample pages for angles are from a Grade 3 book. Students are asked to identify particular angles in polygons. At this point they are asked to decide if an angle is a right angle, less than a right angle, or greater than a right angle. The sample pages for lines are from a Grade 4 book. Intersecting, parallel, and perpendicular lines are emphasized. Students are asked to identify these types of lines.

The *Focus on Skills* problems encourage you to recognize various lines and angles. You will identify real-life objects that can be described or modeled with these geometric shapes. You will demonstrate your understanding of intersecting and parallel lines by identifying the various types of angles that they form.

The *Focus on Concepts* problems encourage you to use your knowledge of these lines and angles to justify other mathematical statements. You will find the measures of specific angles in diagrams involving parallel and intersecting lines and explain the relationships among these angles. Also, you will justify statements about parallelograms and rectangles using your knowledge of parallel lines and the angles they form.

**Chapter 8**
**Lesson**
**4**

# Exploring Angles

**Problem Solving Connection**

Use Objects/ Act It Out

**Materials**

Power Polygons

**Vocabulary**

**polygon**
a closed figure with three or more sides made up of line segments

**angle**
formed by two rays or two line segments with a common end point

**right angle**
an angle that forms a square corner

**Math Tip**

Different polygons may have the same size angle.

**Explore** • • • • • • • • • • • • • • • • • • • • • • • • • • • • • • • • • •

This is a **polygon**. Each corner of a polygon forms an **angle**. In any polygon, the number of sides equals the number of angles.

Angle

Angle

Angle

## Work Together

Use these Power Polygon pieces.

C    G    J    L    I    A

1. Record which pieces have angles that match each.

   a.        b.        c.        d.

2. Find a shape with no angles that are the same.

3. Find a shape with two angles that are the same.

4. Find a shape with all angles the same.

**Talk About It**

How did you decide which pieces had angles that matched **a–d**?

## Connect

Angles can be different sizes.

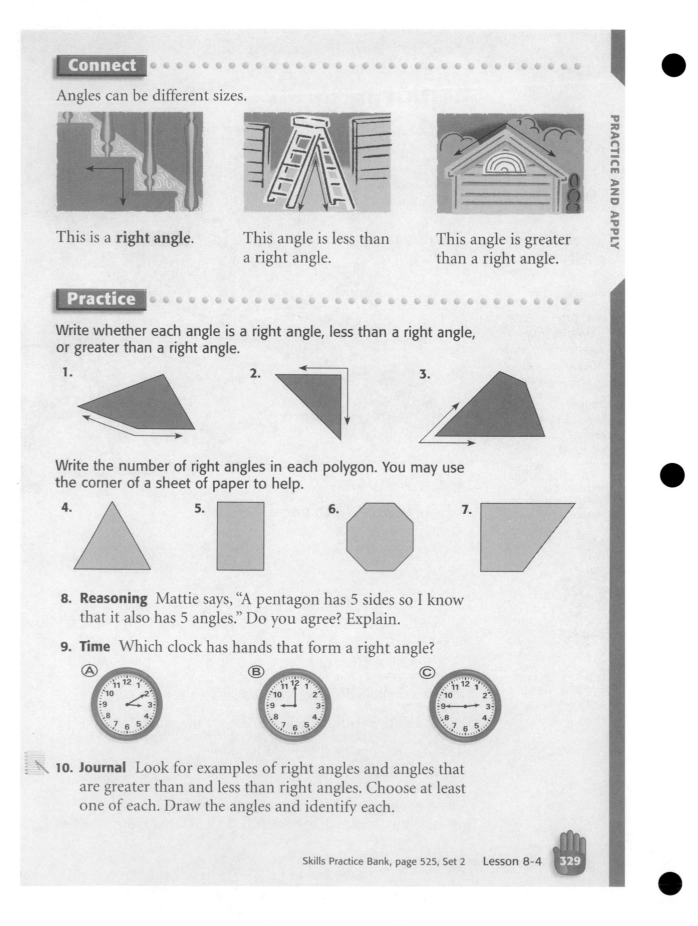

This is a **right angle**.

This angle is less than a right angle.

This angle is greater than a right angle.

## Practice

Write whether each angle is a right angle, less than a right angle, or greater than a right angle.

1.

2.

3.

Write the number of right angles in each polygon. You may use the corner of a sheet of paper to help.

4.

5.

6.

7.

8. **Reasoning** Mattie says, "A pentagon has 5 sides so I know that it also has 5 angles." Do you agree? Explain.

9. **Time** Which clock has hands that form a right angle?

Ⓐ    Ⓑ    Ⓒ

10. **Journal** Look for examples of right angles and angles that are greater than and less than right angles. Choose at least one of each. Draw the angles and identify each.

# Lines and Line Segments

## You Will Learn
how to identify intersecting, parallel, and perpendicular lines

## Vocabulary
**line**
a straight path that goes on forever in both directions

**line segment**
part of a line with two endpoints

**point**
a location in space

**Lines or Line Segments**
intersecting
parallel
perpendicular

## Learn ● ● ● ● ● ● ● ● ● ● ● ● ● ● ● ● ● ● ● ● ● ● ● ● ● ●

Artists use grids. The intersecting lines or line segments help them to enlarge or shrink drawings.

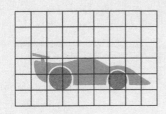

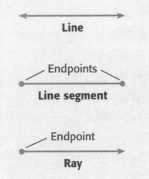

Line

Endpoints

**Line segment**

Endpoint

**Ray**

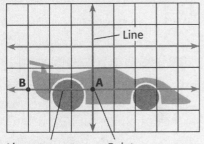

Line

B          A

Line segment          Point
Point A to Point B

In the grid two **intersecting** lines cross at **point** *A*.

**Parallel** lines never intersect. The red grid lines are parallel.

**Perpendicular** lines intersect at right angles. The blue grid line is perpendicular to each red line.

## Talk About It

## Remember
A ray is part of a line with only one end point.

Name pairs of intersecting, perpendicular, and parallel lines or line segments in your classroom.

## Check ● ● ● ● ● ● ● ● ● ● ● ● ● ● ● ● ● ● ● ● ● ● ● ● ● ●

Write intersecting, parallel, or perpendicular for each.

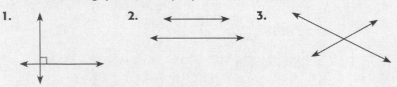

1.          2.          3.

4. **Reasoning** Raoul says perpendicular lines are always intersecting lines. Is he right? Explain.

## Practice

### Skills and Reasoning

Write intersecting, parallel, or perpendicular for each.

5.

6.

7.

8. Write a capital letter of the alphabet that has:

   a. intersecting lines.    b. perpendicular lines.    c. parallel lines.

### Problem Solving and Applications

9. Corey is 1 mile from Stoney Creek. He wants to get to the creek as quickly as possible. Should he walk on a line that is parallel to the creek, perpendicular to the creek, or neither? Explain.

**Using Data** Use the map of Fort Bragg to answer 10 and 11.

10. **Geography** Write if each pair of streets appears to be parallel or perpendicular.

    a. Pine Ave. and Perkins Way

    b. Alder St. and Madrone Ave.

11. **Critical Thinking** Could you get from the corner of McPherson and Laurel to the corner of Fir and Perkins Way without crossing Harrison? Explain.

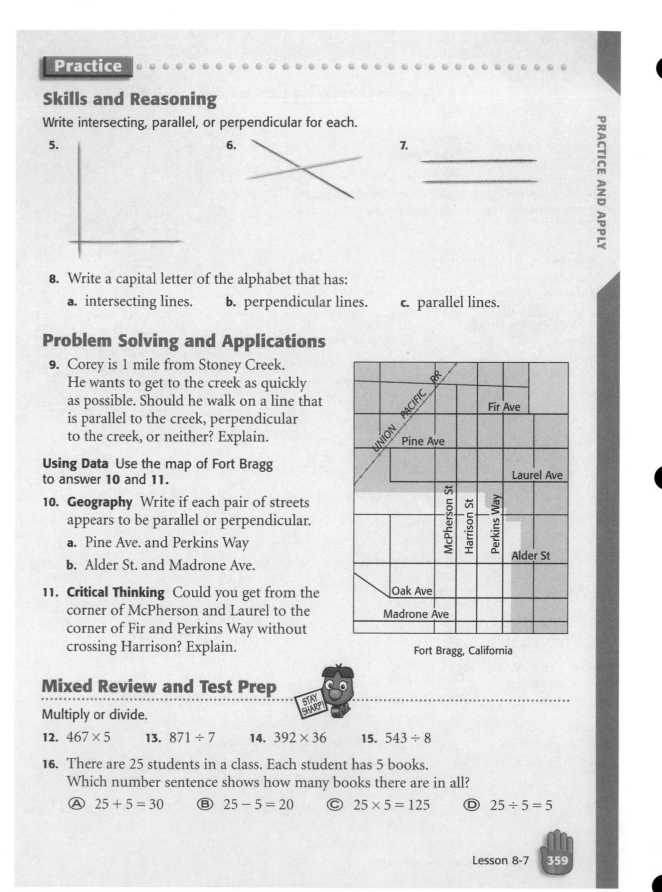

Fort Bragg, California

### Mixed Review and Test Prep

Multiply or divide.

12. $467 \times 5$    13. $871 \div 7$    14. $392 \times 36$    15. $543 \div 8$

16. There are 25 students in a class. Each student has 5 books. Which number sentence shows how many books there are in all?

    Ⓐ $25 + 5 = 30$    Ⓑ $25 - 5 = 20$    Ⓒ $25 \times 5 = 125$    Ⓓ $25 \div 5 = 5$

**FOCUS ON SKILLS**                                  **Name**_____
**Lines and Angles**

1.  List at least two objects or situations in your surroundings that can be described by the following geometric terms.

    a.  line                                    b.  line segment

    c.  ray                                     d.  parallel lines

    e.  perpendicular lines                     f.  skew lines

    g.  acute angle                             h.  right angle

    i.  obtuse angle                            j.  straight angle

2.  Given line *p* intersecting line *q*.

    a.  Identify the pairs of vertical angles.

    b.  Identify the angles forming a linear pair.

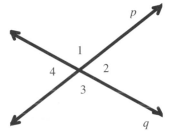

3.  Given line *m* parallel to line *n*.  Find the measures of the angles numbered 1 – 7. Identify the types of angles formed in the diagram.

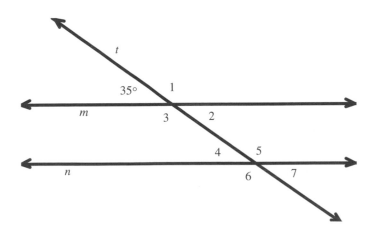

**FOCUS ON CONCEPTS**                          Name_____
**Lines and Angles**

1.  Given line *m* parallel to line *n* with transversals *t* and *r*.  Find the measure of
    angle 1.  Explain your reasoning.

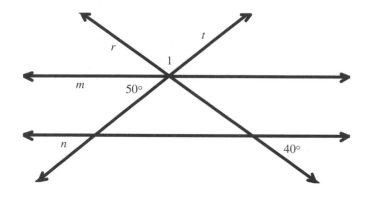

2.  Given line *m* ∥ *n* with transversals *t* and *r*.  How are ∠1, ∠2, and ∠3 related?
    Explain your reasoning.

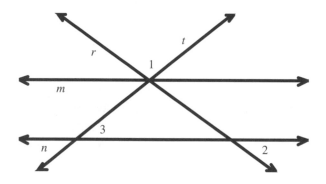

3.  Use your knowledge of parallel lines cut by a transversal and the angles formed
    by them to justify the following statements.

    a.  Consecutive angles in a parallelogram are supplementary.

    b.  The opposite angles of a parallelogram are congruent.

# TRIANGLES

As early as kindergarten, students are encouraged to recognize, name, draw, and construct basic geometric shapes. Among these shapes is the triangle. Students learn to sort triangles by shape and size and recognize these shapes in the environment. As they progress through the elementary grades they classify these triangles according to characteristics such as the measure of the sides and the types of angles. Eventually, they "discover" various properties, axioms, and theorems associated with triangles (e.g. the Pythagorean Theorem).

The sample pages are from a Grade 5 book. The activities on these pages have students classify a triangle by the measure of the sides and by the measure of the angles of the triangle. Also, students are encouraged to notice triangles in the environment and to classify those triangles using the characteristics of sides and angles.

The *Focus on Skills* problems encourage you to become proficient at classifying triangles by the measure of the sides and the measure of the angles. You will measure the sides and the angles of triangles of various shape and size and classify each triangle by sides and angles.

The *Focus on Concepts* problems encourage you to go beyond the skill of classifying triangles according to sides and angles and to develop a better conceptual understanding of the properties of these triangles and the various theorems associated with them. You are asked to think about the characteristics of a particular triangle. For instance, make a list of **all** of the characteristics of a right triangle. Our goal is for students to say more than the fact that a right triangle has a right angle. One of the many concepts we want them to understand is that a right triangle has exactly one right angle and that the other two angles must be acute. Also, we want them to be able to explain why a right triangle can be scalene or isosceles, but not equilateral.

As future teachers of elementary and middle school students, you must be able to do more than the skills that you will teach your students. You must be able to move to a higher level of thinking about the problems and must encourage your students to move to this higher level as well.

# Triangles

## You Will Learn
how to classify
triangles

## Vocabulary
**polygon**
a plane, closed figure
formed by line
segments

**line segment**
a part of a line that
has two end points

**Triangles Classified
by the Measure of
Their Sides**
equilateral
isosceles
scalene

**Triangles Classified
by the Measure of
Their Angles**
right
acute
obtuse

## Learn

Triangles in space! Greg, from Everett,
Washington, just visited the Space Needle
in Seattle with his class. When he described
this structure to a friend, he said he could
see many triangles in it.

Triangles are three-sided **polygons**. Each
side is a **line segment**. They are named
by their vertices. Triangle *ABC* is
formed with line segments
$\overline{AB}$, $\overline{BC}$, and $\overline{CA}$.

You can classify triangles by
the lengths of their sides.

In 2000, approximately 1.3 million visitors
toured the 607 ft tall Space Needle, which
was built for the 1962 World's Fair.

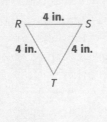

**Equilateral** triangle
all sides the same length

**Isosceles** triangle
at least two sides
the same length

**Scalene** triangle
no sides the
same length

## Remember
An open figure
does not have
all sides
connected.

You can also classify triangles by the measures of their angles.

**Right** triangle
one right angle

**Acute** triangle
all angles less
than 90°

**Obtuse** triangle
one angle greater
than 90°

### Talk About It
Could a triangle be both right and isosceles? Explain.

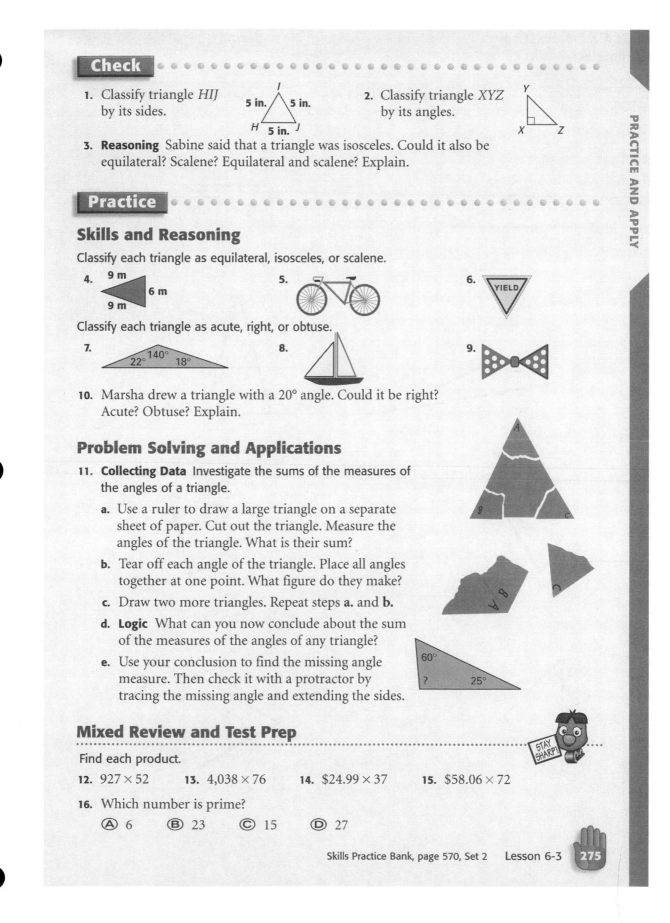

**FOCUS ON SKILLS**                          **Name**_____
**Triangles**

Classify each triangle with as many names as possible using the measurement of the sides and the measurement of the angles. Measure each side and each angle if necessary.

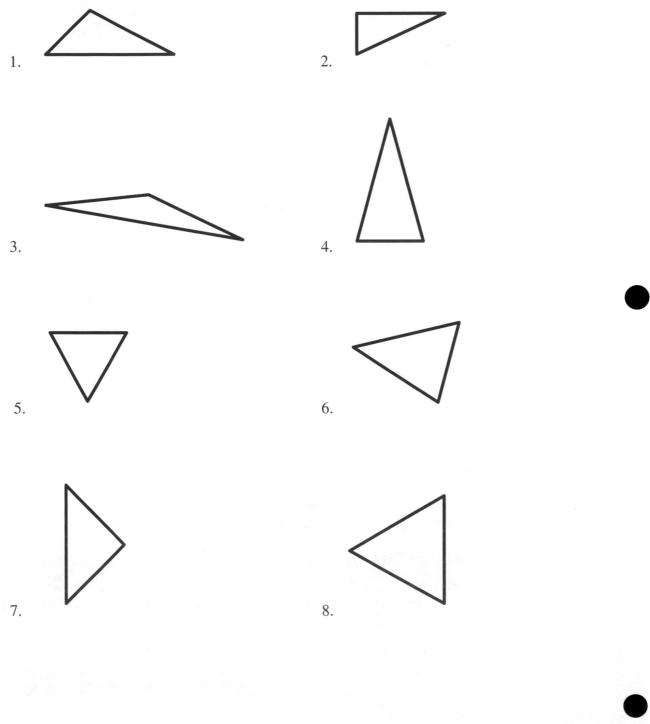

1.

2.

3.

4.

5.

6.

7.

8.

**FOCUS ON CONCEPTS**                              **Name**_____
**Triangles**

You may use the triangles on the *Skills* sheet to help you answer the following questions.

Focus on sides:

1. Is it possible to draw a triangle that is scalene and isosceles? Why or why not?

2. Is it possible to draw a triangle that is scalene and equilateral? Why or why not?

3. Is it possible to draw a triangle that is isosceles and equilateral? Why or why not?

4. Is it possible to draw a triangle with sides 2 cm, 4 cm, and 5 cm? 2 cm, 2cm, and 3 cm? 1 cm, 1cm, and 2 cm? 2 cm, 3 cm, and 6 cm? Why or why not?

5. Refer to the skills sheet or draw several triangles of various size and shape. Measure the sides in each triangle and compare the three sides in each triangle. What relationship exists among the three sides in a triangle?

Focus on angles:

6. Is it possible to draw a triangle that is acute and right? Why or why not?

7. Is it possible to draw a triangle that is acute and obtuse? Why or why not?

8. Is it possible to draw a triangle that is right and obtuse? Why or why not?

9. Refer to the skills sheet or draw several triangles of various size and shape. Measure the angles in each triangle and find the sum of the angles in each triangle. What relationship exists among the three angles in a triangle?

10. Refer to the skills sheet and draw several more right triangles. Describe the three angles in each triangle. What <u>must</u> be true about the angles? What relationship exists among the three angles of a right triangle?

11. Refer to the skills sheet and draw several more equilateral triangles. Describe the three angles in each triangle. What <u>must</u> be true about the angles? What relationship exists among the three angles of an equilateral triangle?

Focus on sides and angles:

12. Is it possible to draw an equilateral, right triangle? Why or why not?

13. Is it possible to draw an equilateral, obtuse triangle? Why or why not?

# QUADRILATERALS

Quadrilaterals are taught at all grade levels in elementary and middle school. Initially, students recognize quadrilaterals by their shape. For instance, kindergarten students recognize squares by their shape. As they progress through the curriculum, students identify other characteristics associated with these shapes such as the measures of the sides and angles. Eventually, properties associated with the diagonals of these shapes are discussed. By Grade 5 students can identify particular quadrilaterals and list their properties.

The sample pages are from a Grade 4 book. The activities on these pages ask students to name quadrilaterals by the lengths of their sides and the sizes of their angles. Also, students are encouraged to compare and contrast the properties of various quadrilaterals and to determine the relationships among them.

The *Focus on Skills* exercises encourage you to become proficient at naming quadrilaterals using the measures of their sides and angles. You will measure the sides and angles of each figure and determine the most appropriate name for it. Also, you will identify all names that are associated with each figure.

The *Focus on Concepts* activities encourage you to go beyond the skill of naming quadrilaterals and to develop a better conceptual understanding of the various types and the relationships among them. You are asked to "make sense" of the relationships among the various quadrilaterals by completing a chart representing these connections. Also, using your own words, you are asked to demonstrate your knowledge of the properties by organizing them in a table. Finally, you are asked to demonstrate your understanding of these properties and relationships by answering "sometimes/always/never" questions.

Many students can (and do) memorize definitions and properties without truly understanding what they mean. Not only should you be able to state a definition, but you must be able to explain it to students and help them understand the connections among the various types of quadrilaterals.

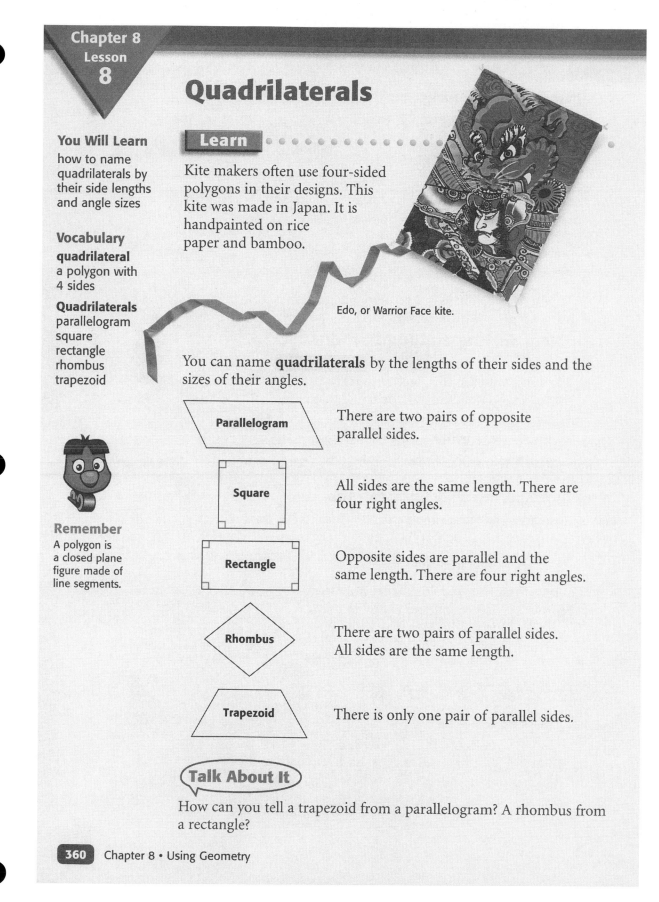

**Chapter 8
Lesson
8**

# Quadrilaterals

**You Will Learn**
how to name quadrilaterals by their side lengths and angle sizes

**Vocabulary**
**quadrilateral**
a polygon with 4 sides

**Quadrilaterals**
parallelogram
square
rectangle
rhombus
trapezoid

**Remember**
A polygon is a closed plane figure made of line segments.

**Learn** • • • • • • • • • • • • • • • • • • •

Kite makers often use four-sided polygons in their designs. This kite was made in Japan. It is handpainted on rice paper and bamboo.

Edo, or Warrior Face kite.

You can name **quadrilaterals** by the lengths of their sides and the sizes of their angles.

**Parallelogram**    There are two pairs of opposite parallel sides.

**Square**    All sides are the same length. There are four right angles.

**Rectangle**    Opposite sides are parallel and the same length. There are four right angles.

**Rhombus**    There are two pairs of parallel sides. All sides are the same length.

**Trapezoid**    There is only one pair of parallel sides.

**Talk About It**

How can you tell a trapezoid from a parallelogram? A rhombus from a rectangle?

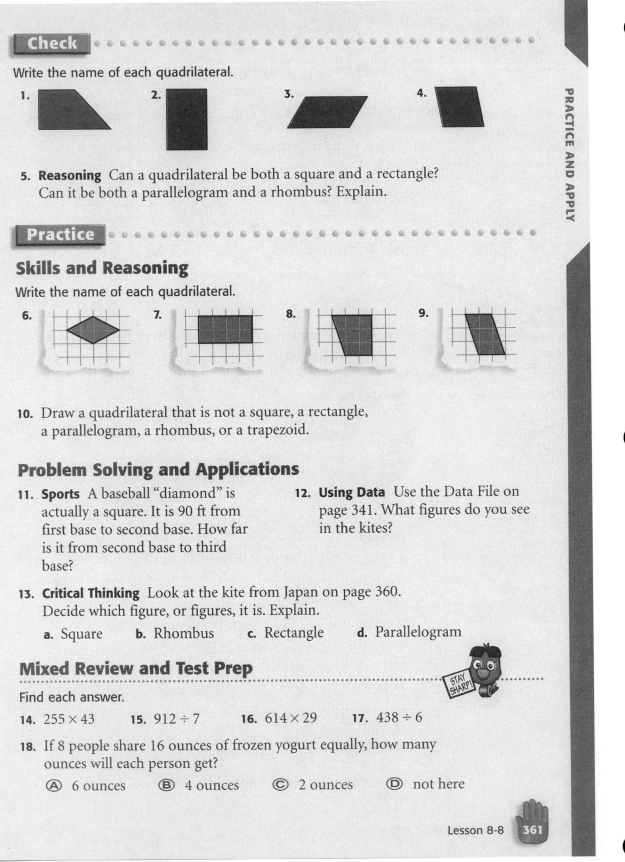

**Check** • • • • • • • • • • • • • • • • • • • • • • • • • • • • • • • • •

Write the name of each quadrilateral.

1.

2.

3.

4.

5. **Reasoning** Can a quadrilateral be both a square and a rectangle? Can it be both a parallelogram and a rhombus? Explain.

**Practice** • • • • • • • • • • • • • • • • • • • • • • • • • • • • • •

## Skills and Reasoning

Write the name of each quadrilateral.

6.

7.

8.

9.

10. Draw a quadrilateral that is not a square, a rectangle, a parallelogram, a rhombus, or a trapezoid.

## Problem Solving and Applications

11. **Sports** A baseball "diamond" is actually a square. It is 90 ft from first base to second base. How far is it from second base to third base?

12. **Using Data** Use the Data File on page 341. What figures do you see in the kites?

13. **Critical Thinking** Look at the kite from Japan on page 360. Decide which figure, or figures, it is. Explain.

    a. Square    b. Rhombus    c. Rectangle    d. Parallelogram

## Mixed Review and Test Prep

Find each answer.

14. $255 \times 43$    15. $912 \div 7$    16. $614 \times 29$    17. $438 \div 6$

18. If 8 people share 16 ounces of frozen yogurt equally, how many ounces will each person get?

    Ⓐ 6 ounces    Ⓑ 4 ounces    Ⓒ 2 ounces    Ⓓ not here

PRACTICE AND APPLY

**FOCUS ON SKILLS**                              **Name**_____
**Quadrilaterals**

A) Name each figure using the words from the list: *quadrilateral, kite, trapezoid, parallelogram, rhombus, rectangle, square.* A figure may have more than one name. Measure the sides and angles, if necessary.

B) What name is the most appropriate for each figure?

1.                                               2.

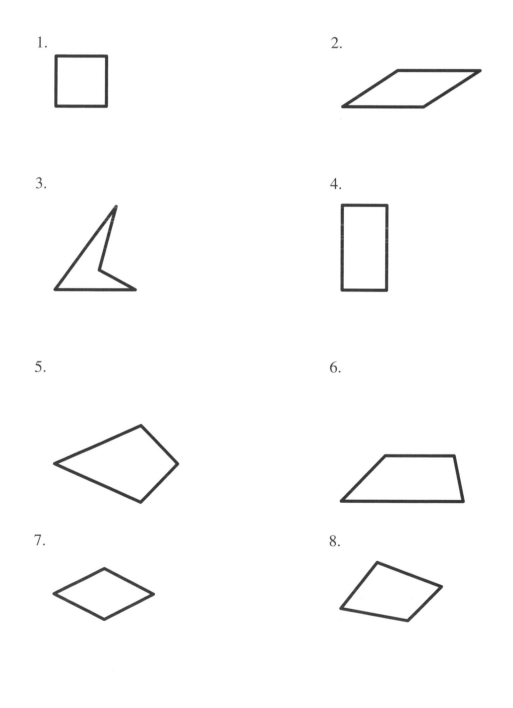

3.                                               4.

5.                                               6.

7.                                               8.

**FOCUS ON CONCEPTS**                                     **Name**_____
**Quadrilaterals**

Place the following words in the chart to demonstrate the relationships among the various types of quadrilaterals.

Word List:  Kite, Parallelogram, Rectangle, Rhombus, Square, Trapezoid

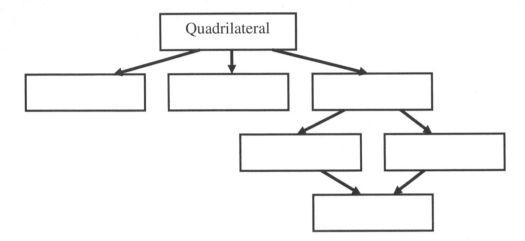

Make a table similar to the one below and list the properties of each type of quadrilateral under the appropriate heading.  Include information about sides, angles, and diagonals in the list.

| Quadrilateral | Kite | Trapezoid | Parallelogram | Rectangle | Rhombus | Square |
|---|---|---|---|---|---|---|
|  |  |  |  |  |  |  |

Write *sometimes*, *always*, or *never* in the blank by each statement.

_____ 1.  A square is a rectangle.

_____ 2.  A rectangle is a rhombus.

_____ 3.  A parallelogram is a square.

_____ 4.  A trapezoid is a kite.

_____ 5.  A rhombus is a square.

_____ 6.  A kite is a quadrilateral.

_____ 7.  The diagonals of a trapezoid are congruent.

_____ 8.  The diagonals of a kite bisect each other.

_____ 9.  The diagonals of a rectangle are congruent.

_____10.  The diagonals of a parallelogram are perpendicular bisectors.

## POLYGONS

Polygons are found throughout the elementary and middle school curriculum.  They are two-dimensional shapes or closed plane figures formed by the union of line segments.  At a very young age, children learn to recognize some of these shapes.  Before entering kindergarten most children can identify shapes such as squares and triangles.  In kindergarten, polygons are referred to as shapes or figures.  By third grade, students are counting the number of sides and corners (vertices).  As they progress through the grade levels they learn to name polygons, recognize regular polygons, and explore relationships among the angles in a polygon.

The sample pages are from a Grade 4 book.  The activities on these pages ask students to name various polygons.  Also, students must determine the number of sides of several polygons and recognize that the sides of a polygon do not have to be equal.

The *Focus on Skills* problems refer to regular polygons.  You are encouraged to become proficient at finding the number of sides, the number of angles, the sum of the measures of the interior angles, the measure of an interior angle, the measure of an exterior angle, the sum of the measures of the exterior angles, and the measure of the central angle for specific polygons.  Also, you will draw several of these polygons using a ruler, a protractor, and the values you determined.

The *Focus on Concepts* questions encourage you to look for patterns in the list of numbers generated on the *Skills* sheet and to develop a "rule", formula, or strategy for finding the various measures for a polygon with *n* sides.  Students may identify several patterns, but ultimately we want them to generalize these patterns so they will be able to find measures for a polygon with a large number of sides easily.  For example, how would you find the number of diagonals or the measure of an interior angle of a regular 100-gon?  Our goal is not for students to memorize a formula, but to be able to re-create a formula or rule for solving problems.

# Exploring Polygons

**Problem Solving Connection**
Make a Table

**Materials**
ruler

**Vocabulary**
**plane figure**
a figure that lies on a flat surface

**polygon**
a closed plane figure made of line segments

**Polygons**
triangle
quadrilateral
pentagon
hexagon
octagon

**Explore** • • • • • • • • • • • • • • • • • • • • • • • • •

Inca sculptors in Peru built walls out of huge rocks. They carved the rocks so perfectly that they fit together with no space between them.

Stones used in the terraces of the Inca fort of Sacsahuaman are almost polygon-shaped.

## Work Together

**Did You Know?**
Each of the four triangular faces of the Great Pyramid at Giza, Egypt, were once 481 feet high, about as high as a 40-story building.

1. Study the 3 figures outlined in the photo. Record each figure's letter and the number of sides it has.

| Outlined Figures | Number of Sides | Other Figures |
|---|---|---|
| A | 4 | K |
| B | | |
| C | | |

2. For each outlined figure, find other figures in the photo with the same number of sides. Record their letters.

3. Use a ruler to draw a wall with five blocks. Draw blocks with 3 sides, 4 sides, 5 sides, 6 sides, and 8 sides.

**Talk About It**

Look at some figures in the photo with the same number of sides. How are the figures different?

## Connect

The face of a solid is a **plane figure**. A **polygon** is a closed plane figure formed by line segments. Here are some polygons.

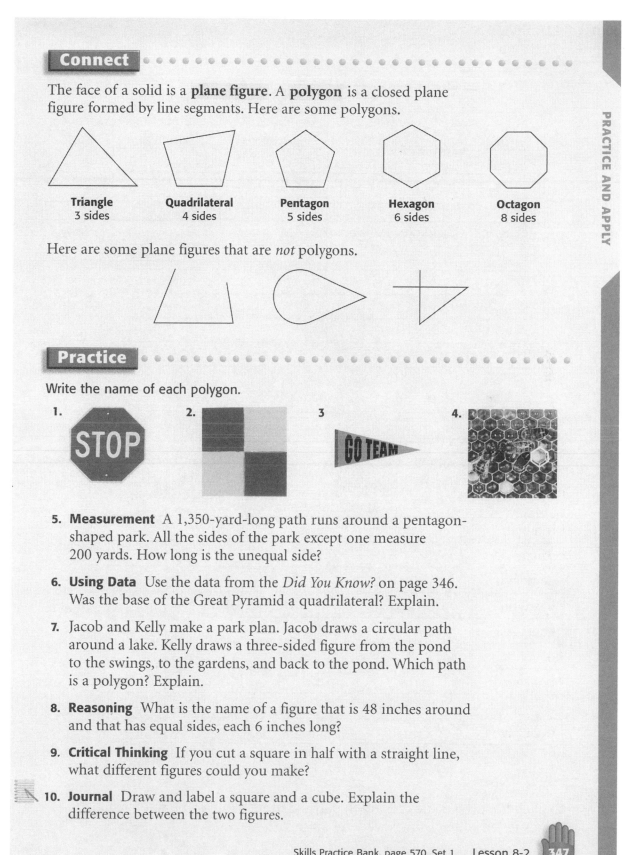

**Triangle**
3 sides

**Quadrilateral**
4 sides

**Pentagon**
5 sides

**Hexagon**
6 sides

**Octagon**
8 sides

Here are some plane figures that are *not* polygons.

## Practice

Write the name of each polygon.

1. STOP
2. 
3. GO TEAM
4. 

5. **Measurement** A 1,350-yard-long path runs around a pentagon-shaped park. All the sides of the park except one measure 200 yards. How long is the unequal side?

6. **Using Data** Use the data from the *Did You Know?* on page 346. Was the base of the Great Pyramid a quadrilateral? Explain.

7. Jacob and Kelly make a park plan. Jacob draws a circular path around a lake. Kelly draws a three-sided figure from the pond to the swings, to the gardens, and back to the pond. Which path is a polygon? Explain.

8. **Reasoning** What is the name of a figure that is 48 inches around and that has equal sides, each 6 inches long?

9. **Critical Thinking** If you cut a square in half with a straight line, what different figures could you make?

10. **Journal** Draw and label a square and a cube. Explain the difference between the two figures.

**FOCUS ON SKILLS**
**Regular Polygons**

Name_____

1.  Complete the table.

| Polygon | number of sides | number of vertices | number of diagonals |
|---------|------|------|------|
| triangle | | | |
| square | | | |
| pentagon | | | |
| hexagon | | | |
| heptagon | | | |
| octagon | | | |
| nonagon | | | |
| decagon | | | |
| dodecagon | | | |

2.  Complete the table.

| Polygon | sum of interior angles | measure of interior angle | measure of exterior angle | sum of exterior angles | measure of central angle |
|---------|------|------|------|------|------|
| triangle | | | | | |
| square | | | | | |
| pentagon | | | | | |
| hexagon | | | | | |
| heptagon | | | | | |
| octagon | | | | | |
| nonagon | | | | | |
| decagon | | | | | |
| dodecagon | | | | | |

3.  Explain how you could draw each polygon using a ruler, a protractor, and the values in the table.  Draw several of these.

**FOCUS ON CONCEPTS**                          Name_____
**Regular Polygons**

1. Refer to the *Skills* sheet and examine the number of diagonals for each polygon. Identify any patterns in the list of numbers.

2. How many diagonals does a polygon with *n* sides (*n*-gon) have?  Explain.

3. Look for patterns and relationships in the other lists of numbers on the *Skills* sheet. Determine a "rule" for finding each of the following for a *n*-gon.

   - Sum of the interior angles

   - Measure of an interior angle

   - Measure of an exterior angle

   - Sum of measures of the exterior angles

   - Measure of a central angle

4. Will your "rule" for the sum of the measures of the interior angles work for any polygon (not just regular)?  Explain.

5. Will your "rule" for the measure of an interior angle work for any polygon (not just regular)?  Explain.

## POLYHEDRA

Polyhedra are three-dimensional solids whose faces are polygons. These shapes define much of our everyday lives. Toasters, microwaves, and refrigerators are rectangular prisms. Also, many buildings form a rectangular prism. However, some buildings resemble other prisms (such as the pentagon), as well as pyramids and cylinders.

As with particular polygons, children learn to identify certain polyhedra at an early age. Cubes and blocks (rectangular prisms) are familiar items to most kindergarten students. Also, other solids with curved surfaces such as balls (spheres), cans (cylinders), and cones are recognizable. The study of solid figures and their properties extends throughout the mathematics curriculum.

The sample pages are from a Grade 4 book. These pages illustrate the difference between polyhedra and solids with curved surfaces. Students are asked to name specific solids and the shapes forming the solids. Also, they are asked to count the number of faces, edges, and vertices of solids with flat surfaces (i. e. polyhedra).

The *Focus on Skills* activity identifies the five platonic solids, various prisms, and various pyramids. You will identify the faces of the solids (i.e. triangles, squares, rectangles, etc.) and count the number of vertices, faces, and edges of each. It will be beneficial to you to have models of each solid or at least a picture of each.

The *Focus on Concepts* questions encourage you to go beyond counting and explore the relationship among the vertices, faces, and edges of polyhedra. Also, you will develop a "rule", formula, or strategy for determining the number of vertices, faces, and edges for an *n*-gon prism and *n*-gon pyramid. For instance, can you find these values for a prism with a base that has 100 sides? Our goal is not for students to memorize a formula, but to be able to re-create a formula or rule for solving problems.

**Chapter 8
Lesson
1**

# Exploring Solids

**Problem Solving Connection**
Use Objects/
Act It Out

**Vocabulary**

**Solid Figures**
cube
rectangular prism
pyramid
cone
sphere
cylinder

**face**
If a solid figure has only flat surfaces, then we call each surface a face.

**edge**
a line segment where two faces of a solid meet

**vertex**
a point where two or more edges meet

**Explore** • • • • • • • • • • • • • • • • • • • • • • • • • • • •

More than 4,000 years ago, sculptors carved huge blocks of stone at Stonehenge in England. Some blocks weigh about 50 tons. Each stone block is a **solid figure**. Many solids have only flat surfaces. They are made up of **faces**, **edges**, and **vertices**.

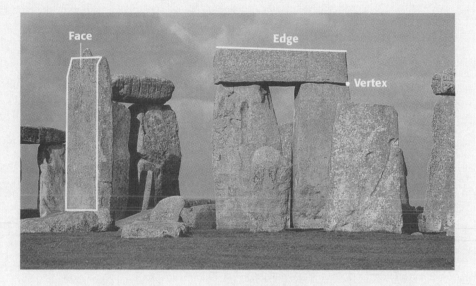

## Work Together

1. Find four solids that have only flat surfaces in your classroom. Trace and label each face on paper.

2. Copy and complete the table using your four solids.

| Solid | Number of Faces | Number of Edges | Number of Vertices |
|-------|-----------------|-----------------|--------------------|
| Book  | 6               | 12              |                    |
|       |                 |                 |                    |
|       |                 |                 |                    |
|       |                 |                 |                    |

### Talk About It

3. Which of your solids has the most edges? Which has the most vertices?

4. Are any of the faces that you traced alike?

## Connect

You can describe solids in different ways. These solids have only flat surfaces.

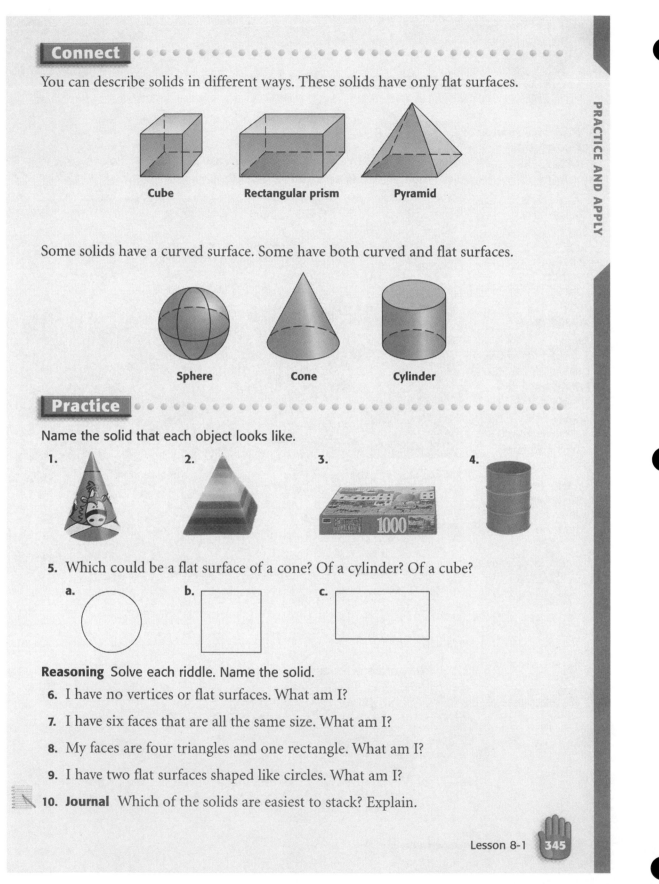

**Cube** **Rectangular prism** **Pyramid**

Some solids have a curved surface. Some have both curved and flat surfaces.

**Sphere** **Cone** **Cylinder**

## Practice

Name the solid that each object looks like.

1.
2.
3.
4.

5. Which could be a flat surface of a cone? Of a cylinder? Of a cube?

a.
b.
c.

**Reasoning** Solve each riddle. Name the solid.

6. I have no vertices or flat surfaces. What am I?

7. I have six faces that are all the same size. What am I?

8. My faces are four triangles and one rectangle. What am I?

9. I have two flat surfaces shaped like circles. What am I?

10. **Journal** Which of the solids are easiest to stack? Explain.

Name_____

**FOCUS ON SKILLS**
**Polyhedra**

Complete the chart:
(Refer to the models or pictures of each solid provided by your instructor.)

| | Name | Identify the faces | Number of Faces (F) | Number of Vertices (V) | Number of Edges (E) |
|---|---|---|---|---|---|
| **PLATONIC Solids** | Tetrahedron | | | | |
| | Cube | | | | |
| | Octahedron | | | | |
| | Dodecahedron | | | | |
| | Icosahedron | | | | |
| **PRISMS** | Triangular Prism | | | | |
| | Rectangular Prism | | | | |
| | Pentagonal Prism | | | | |
| | Hexagonal Prism | | | | |
| **PYRAMIDS** | Triangular Pyramid | | | | |
| | Rectangular Pyramid | | | | |
| | Pentagonal Pyramid | | | | |
| | Hexagonal Pyramid | | | | |

**FOCUS ON CONCEPTS**                              Name_____
**Polyhedra**

1.  What relationship exists among the vertices (V), faces (F), and edges (E) of the
    Platonic solids?

2.  Does this relationship exist for the prisms on the *Skills* sheet? the pyramids?

3.  Describe an *n*-gon prism.

4.  Develop a "rule" for finding the number of vertices, faces, and edges in an *n*-gon
    prism.

5.  Describe an *n*-gon pyramid.

6.  Develop a "rule" for finding the number of vertices, faces, and edges in an *n*-gon
    pyramid.

7.  Does the relationship described in question number one exist for the *n*-gon prism
    and the *n*-gon pyramid?  Explain.

# MEASUREMENT

Measurement is taught in the early elementary grades and is reinforced throughout the mathematics curriculum. By second grade students are measuring objects using nonstandard units, as well as English units. As they progress through the grades, they also learn to use metric units. Measurement is an essential skill. We measure lots of things in our everyday lives (our clothes, shoes, weight, temperature, to name a few). Distance runners measure how many miles they run. Home builders measure lengths of boards. Surveyors measure the size of lots or fields. Meteorologists measure the amount of rainfall. Even when we wrap a present for someone, we measure the wrapping paper to determine the length of the needed piece.

The sample pages are from two elementary school books and include activities involving three types of units: nonstandard, English, and metric. The first set of pages is from a Grade 2 book. The activities on these pages ask students to measure objects (e. g. desk, chair, chalkboard) using nonstandard units of measure (paper clip, pencil, snap cube). The second set of pages is also from the Grade 2 book. Students are asked to measure objects using English measurements such as inches, feet, and yards. The final set of pages is from a Grade 4 book. On these pages students are asked to measure using metric units.

The *Focus on Skills* problems encourage you to improve your measurement skills using all three types of measurements. You will measure objects such as your index finger and your desk with paper clips and Q-tips. Also, you will measure these objects using inches and centimeters as your units of measure. The last two problems on the sheet ask you to measure the chalkboard and classroom door using feet, yard, and meters. These activities provide a comparison between the English measurements (feet and yards) and the metric (meters) ones.

The *Focus on Concepts* problems go beyond the "act of measuring" and encourage you to make connections among the various units within each system. You will convert from one unit to another (e.g. feet to inches) in the English system and explain how you made the conversion. You will make similar conversions and explanations within the metric system. Finally, you will explore connections between the two systems by converting a temperature measured in Fahrenheit to one measured in Celsius and a rate of a car measured in miles per hour to kilometers per hour.

Name _____

# Explore Nonstandard Units

**Explore** • • • • • • • • • • • • • • • • • • • • • • • • • •

Choose an object in your classroom to measure. Measure its length.

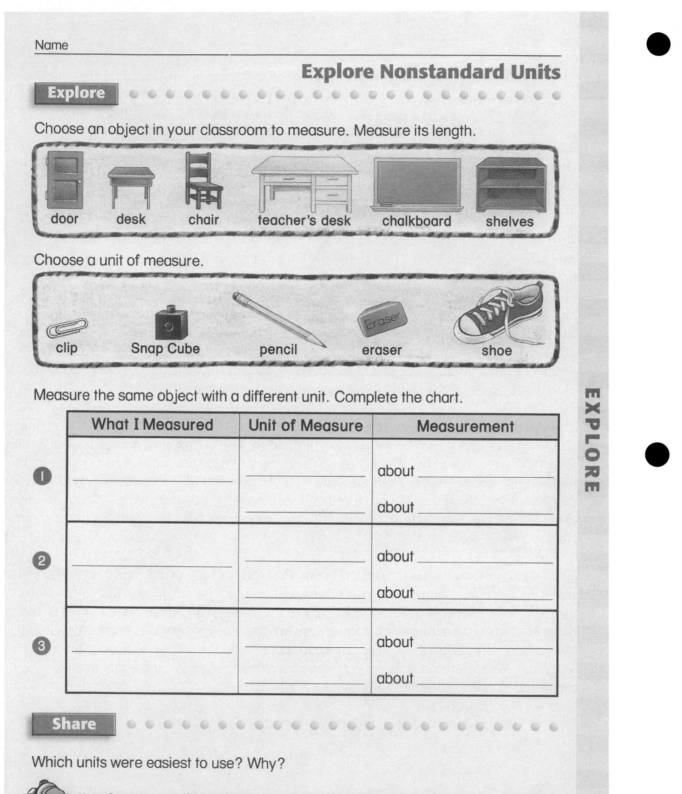

| | | | | | |
|---|---|---|---|---|---|
| door | desk | chair | teacher's desk | chalkboard | shelves |

Choose a unit of measure.

| | | | | |
|---|---|---|---|---|
| clip | Snap Cube | pencil | eraser | shoe |

Measure the same object with a different unit. Complete the chart.

| | What I Measured | Unit of Measure | Measurement |
|---|---|---|---|
| **1** | _____ | _____ | about _____ |
| | | _____ | about _____ |
| **2** | _____ | _____ | about _____ |
| | | _____ | about _____ |
| **3** | _____ | _____ | about _____ |
| | | _____ | about _____ |

**Share** • • • • • • • • • • • • • • • • • • • • • • • • • •

Which units were easiest to use? Why?

**Notes for Home:** Your child used different objects to measure the lengths of objects in the classroom.
*Home Activity:* Ask your child to measure the length of his or her bed with an object such as a spoon or straw.

**Chapter 11 Lesson 1**

**EXPLORE**

Connect • • • • • • • • • • • • • • • • • • • • • • • • • • • •

Estimate the lengths.
Use Snap Cubes to measure.
Write the numbers.

*I think it looks about 5 cubes long.*

**4**

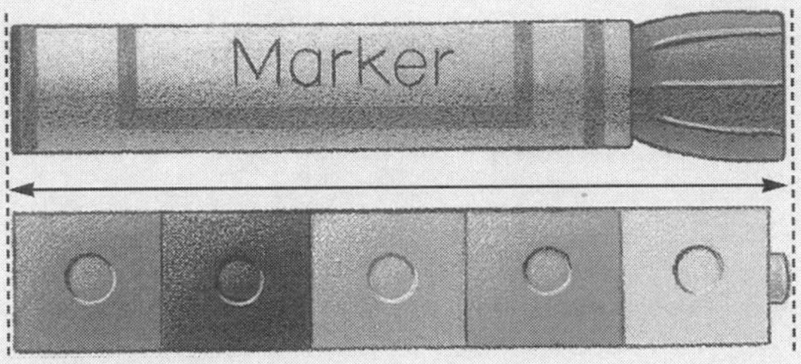

Marker

Estimate: about _____ Snap Cubes          Measure: about _____ Snap Cubes

**5**

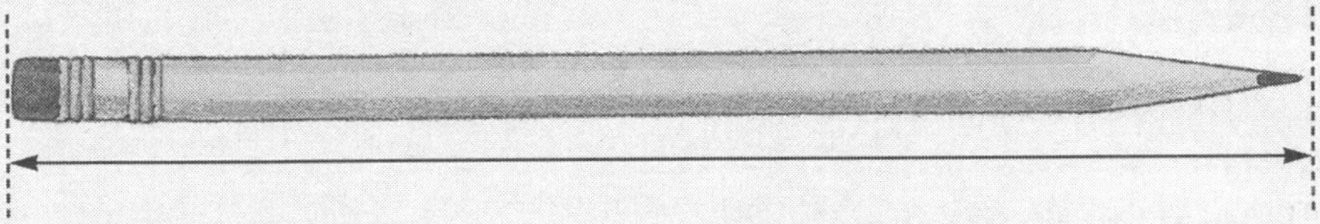

Estimate: about _____ Snap Cubes          Measure: about _____ Snap Cubes

**6**

Glue Stick

Estimate: about _____ Snap Cubes          Measure: about _____ Snap Cubes

## Problem Solving  Critical Thinking

**7** Pablo measured how tall he is.
First he measured with paper clips.
Then he measured with pencils.
Did Pablo need more paper clips
or more pencils? Explain.

**Notes for Home:** Your child estimated and measured the lengths of objects using Snap Cubes.
*Home Activity:* Ask your child to use a paper clip or another similar object to measure the lengths
of objects in the kitchen.

Name _____

# Inches, Feet, and Yards

**Learn** • • • • • • • • • • • • • • • • • • • • • • • • • • • • •

Yuko just completed a jump.
She wants to find out how long it is.
She can measure length in inches,
feet, and yards.

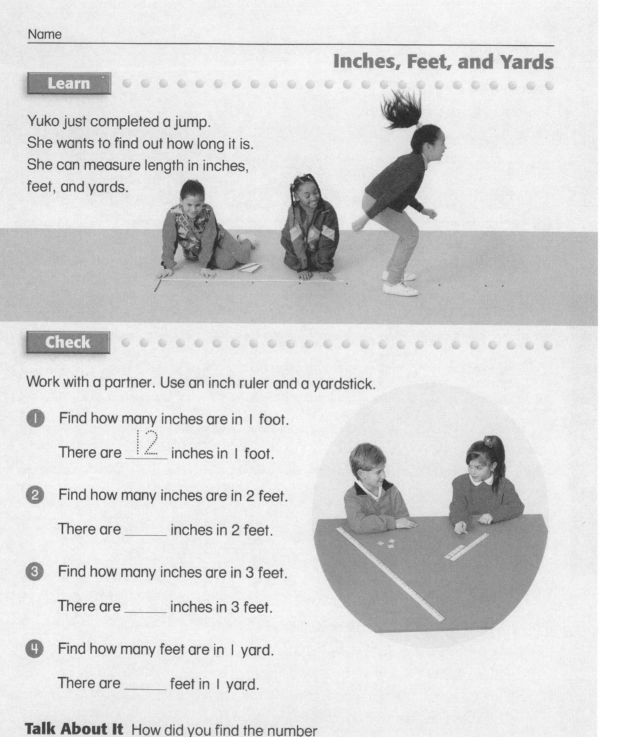

**Check** • • • • • • • • • • • • • • • • • • • • • • • • • • • • •

Work with a partner. Use an inch ruler and a yardstick.

1   Find how many inches are in 1 foot.

There are __12__ inches in 1 foot.

2   Find how many inches are in 2 feet.

There are _____ inches in 2 feet.

3   Find how many inches are in 3 feet.

There are _____ inches in 3 feet.

4   Find how many feet are in 1 yard.

There are _____ feet in 1 yard.

**Talk About It** How did you find the number
of inches in 1 foot? the number of feet in 1 yard?

**Notes for Home:** Your child investigated inches, feet, and yards. *Home Activity:* Ask your child to tell
you which is longer—an inch, a foot, or a yard. (yard)

**Practice** • • • • • • • • • • • • • • • • • • • • • • • • • • •

Complete the chart.

| Estimate. Find an object about this long. | Write or draw your object. | Measure length to check. |
|---|---|---|
| ⑤ about 1 inch | | about _____ |
| ⑥ about 1 foot | | about _____ |
| ⑦ about 1 yard | | about _____ |

## Problem Solving Estimation

Circle the best estimate for the length of each object.

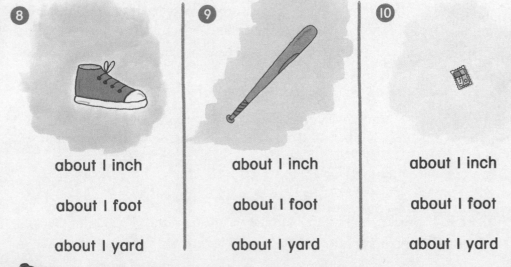

⑧

about 1 inch

about 1 foot

about 1 yard

⑨

about 1 inch

about 1 foot

about 1 yard

⑩

about 1 inch

about 1 foot

about 1 yard

**Notes for Home:** Your child found and measured objects that were about 1 inch, 1 foot, and 1 yard long. *Home Activity:* Ask your child to find objects at home that are about 1 inch, 1 foot, and 1 yard long.

# Exploring Centimeters, Decimeters, and Meters

**Problem Solving Connection**
- Use Objects/ Act It Out
- Guess and Check

**Materials**
- string
- felt-tip marker
- meter stick
- baseball, tennis ball, and other sports equipment

**Vocabulary**
**Metric Units of Length**
centimeter
decimeter
meter

**Math Tip**
The prefix *deci-* (as in *decimal*) means 10. The prefix *cent-* (as in *century*) means 100.

**Explore** • • • • • • • • • • • • •

Handball is a modern sport with ancient roots. A form of handball may have been played in ancient Greece as early as 600 B.C. It became an Olympic sport for the first time in 1936.

## Work Together

The distance around a handball is 56 centimeters. You can use this measurement to help you estimate other short distances.

In the 2000 Summer Olympics, the Russian Federation won the gold medal in men's handball.

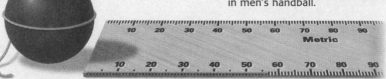

1. Estimate the distance around a baseball. Do you think it is more or less than the distance around a handball?
   a. Wrap string around a baseball and cut it so the ends meet.
   b. Place the string on a meter stick. Measure the length of the string. Record your measurement.

2. Find other sports balls, such as a softball, a tennis ball, a football, and a basketball. Estimate and measure. Record their measurements.

**Talk About It**

3. Which sports balls have a greater distance around than a handball?

4. How can you tell where 56 cm is on the meter stick above?

PRACTICE AND APPLY

## Connect

**Centimeter** (cm), **decimeter** (dm), and
**meter** (m) are metric units of length.

| |
|---|
| 10 cm = 1 dm |
| 10 dm = 1 m |
| 100 cm = 1 m |

Here are some ways to think about centimeters, decimeters, and meters.

A thumbtack is
about 1 cm wide.

A cassette tape is
about 1 dm wide.

A baseball bat is
about 1 m long.

## Practice

Choose the better unit of measure for each object.

**1.**
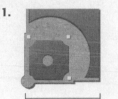

decimeters or meters

**2.**

centimeters or meters

**3.**

decimeters or meters

Copy and complete. Write >, <, or =.

**4.** 12 cm ● 12 dm     **5.** 4 m ● 4 cm     **6.** 50 dm ● 50 m     **7.** 2 m ● 2 dm

▲ **Geometry Readiness** Use a metric ruler to draw each shape.

**8.** A rectangle 1 dm long and 6 cm wide

**9.** A parallelogram with two 3-cm sides
and two 5-cm sides

**10. History** Fencing competitions were
held in Egypt and Japan as long as
5,000 years ago. Today, Olympic
fencers compete in a rectangular
area that is 14 m long and 2 m wide.
What is the perimeter of the rectangle?

**11. Journal** Give examples of when you
would measure in meters, decimeters,
and centimeters.

In the 2000 Summer Olympics, 235 people from
43 countries entered the fencing competitions.

**FOCUS ON SKILLS**                          **Name**_____
**Measurement**

Nonstandard Units:

Measure the following objects using a small paper clip, a large paper clip, and a Q-tip.

    1.    The length of your index finger

    2.    The length of your arm from your wrist to your elbow

    3.    The dimensions of the top of your desk

English Units:

Measure the following objects using a ruler marked in inches.

    4.    The length of your index finger

    5.    The length of your arm from your wrist to your elbow

    6.    The dimensions of the top of your desk

Metric Units:

Measure the following objects using a ruler marked in centimeters.

    7.    The length of your index finger

    8.    The length of your arm from your wrist to your elbow

    9.    The dimensions of the top of your desk

More Measurements:

    10.    Measure the chalkboard in (a) feet, (b) yards, and (c) meters.

    11.    Measure the classroom door in (a) feet, (b) yards, and (c) meters.

**FOCUS ON CONCEPTS**                     Name_____
**Measurement**

Convert the given amounts to the indicated units.  Explain how each pair of units is related.


Conversions within the English system:

1.  27 inches = _____ feet

2.  5 yards = _____ feet

3.  2 miles = _____ inches

4.  32 quarts = _____ gallons

5.  3.6 pounds = _____ ounces


Conversions within the Metric system:

6.  30 meters = _____ centimeters

7.  306 millimeters = _____ meters

8.  5.3 liters = _____ centiliters

9.  540 grams = _____ kilograms

10.  35 millimeters = _____ centimeters


Conversions between English and Metric systems:

11.  85 °Fahrenheit = _____ °Celsius

12.  60 miles per hour = _____ kilometers per hour

## PERIMETER AND AREA

Perimeter and area are taught in the elementary grades as early as second grade. For some shapes, perimeter can be written as a standard formula. However, perimeter can more generally be thought of as the measure "around" a shape. By second grade students learn to measure length using various units. Thus, it is a natural progression to define perimeter as the sum of the lengths of the sides of a shape. In contrast, area is more often thought of in terms of formulas. However, in the elementary grades students develop the concept of area using dot paper and grids.

The sample pages for perimeter are from a Grade 2 book. Students are asked to measure the sides of objects with a ruler and to find the perimeter by adding the lengths of the sides. Also, they are asked to determine which shape has the greatest perimeter based on measurements and on visual examination without measurement.

The sample pages for area are from a Grade 4 book. Students are asked to explore the areas of rectangles by dividing the rectangle into squares and counting the number of squares that cover the rectangle. This task helps students understand the concept of area being measured in "square units." Also, students use the formula $A = l \times w$ to find areas. The "Work Together" activity on these pages is a great activity for approximating the area of an irregular shape. Students trace each others shoe on grid paper and estimate the area of the bottom of the shoe by counting the squares included within the boundary of the drawing. In this exercise students must combine "partial" squares to make complete squares. You can estimate the area of the palm of your hand in a similar manner.

The *Focus on Skills* problems encourage you to practice finding area and perimeter. Some of these problems ask you to find the perimeter and area of various shapes drawn on centimeter dot paper. These problems reinforce the concept of dividing the shape into square units to determine the area. Other problems can be completed using an appropriate formula. Pictures are provided with some of the problems, while others require you to draw the shape, if needed.

The *Focus on Concepts* problems encourage you to explore relationships, as well as distinctions, between perimeter and area. Given a specific perimeter, you will draw all possible rectangles with whole number dimensions and examine the areas of these rectangles. Similarly, given a specific area, you will determine the dimensions and perimeters of possible rectangles. Finally, you will explore "non-rectangular" shapes using a string of a specific length over a peg board.

Name _____

## Perimeter

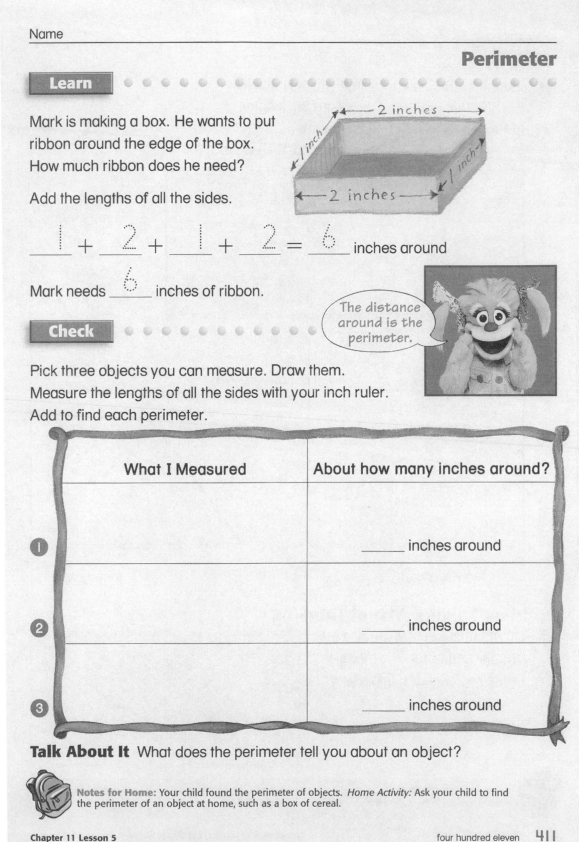

**Learn** • • • • • • • • • • • • • • • • • • • • • • • • • •

Mark is making a box. He wants to put ribbon around the edge of the box. How much ribbon does he need?

2 inches

1 inch

1 inch

2 inches

Add the lengths of all the sides.

__1__ + __2__ + __1__ + __2__ = __6__ inches around

Mark needs __6__ inches of ribbon.

The distance around is the perimeter.

**Check** • • • • • • • • • • • • • • •

Pick three objects you can measure. Draw them.
Measure the lengths of all the sides with your inch ruler.
Add to find each perimeter.

| | What I Measured | About how many inches around? |
|---|---|---|
| 1 | | _____ inches around |
| 2 | | _____ inches around |
| 3 | | _____ inches around |

**Talk About It** What does the perimeter tell you about an object?

**Notes for Home:** Your child found the perimeter of objects. *Home Activity:* Ask your child to find the perimeter of an object at home, such as a box of cereal.

**Practice** • • • • • • • • • • • • • • • • • • • • • • • • • • • • • • • •

4  Mark an **X** on the shape that you estimate has
the greatest perimeter. Measure the lengths
of the sides. Add to find the perimeter.
Circle the shape with the greatest perimeter.

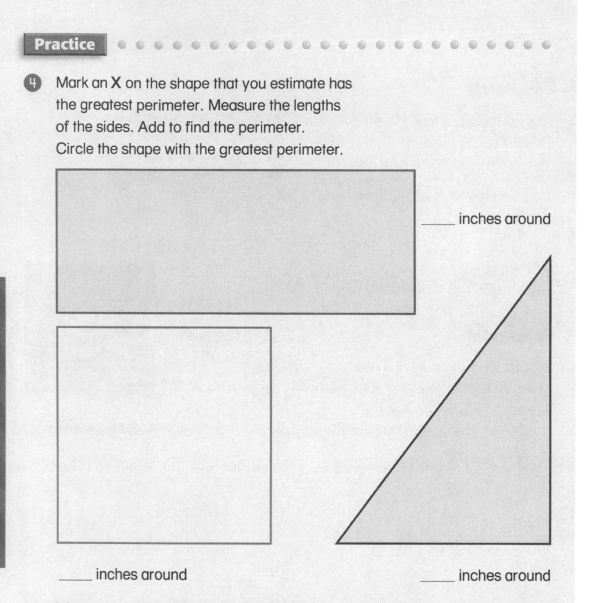

_____ inches around

_____ inches around

_____ inches around

## Problem Solving **Visual Thinking**

5  Do not measure. Which has the
greater perimeter, the triangle
or the rectangle? Explain why.

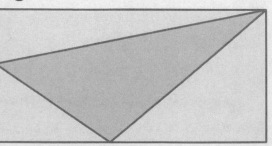

**Notes for Home:** Your child practiced finding the perimeter of different shapes. *Home Activity:* Ask your
child to explain how he or she found each perimeter on the page.

**Chapter 8**
**Lesson**
**12**

# Exploring Areas of Rectangles

**Problem Solving Connection**
Use Objects/
Act It Out

**Materials**
grid paper

**Vocabulary**
**area**
number of square
units needed to
cover a figure

**Some Units for Measuring Area**
square unit
square inch
square centimeter

### Explore ● ● ● ● ● ● ● ● ● ● ● ● ● ● ● ● ● ● ● ● ● ● ● ● ● ● ● ● ●

The **area** of a figure is the number of **square units** needed to cover it.

This is a square unit.      This is about half of a square unit.

## Work Together

Estimate the area of the bottom of your shoe.
Use grid paper.

1. **a.** Trace the outline of your shoe
   on grid paper.

   **b.** Count the whole squares and
   almost-whole squares.

   **c.** Count the half squares and
   almost-half squares.

   **d.** Estimate the total area.

**Math Tip**
To get a closer
estimate, try counting
2 half-squares as
1 whole square.

### Talk About It

2. How did you estimate the total area of your shoe?

3. If someone else estimated the area of your shoe, would they get
   the same answer? Explain.

## Connect

You use square units to measure area. You can count the number of **square centimeters** or **square inches** to find the area.

The area is 5 square centimeters.        The area is 2 square inches.

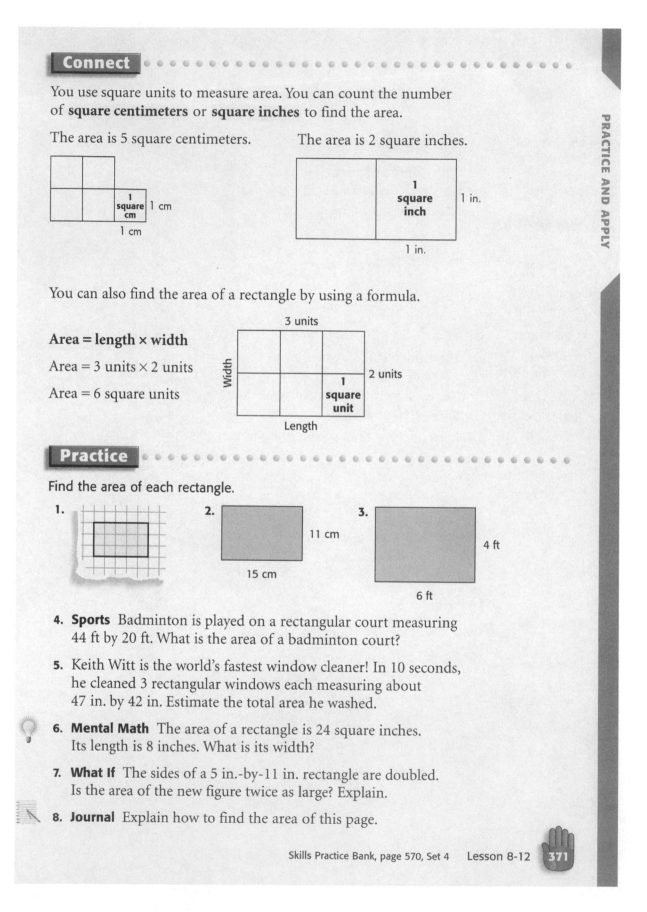

You can also find the area of a rectangle by using a formula.

**Area = length × width**

Area = 3 units × 2 units

Area = 6 square units

## Practice

Find the area of each rectangle.

1.

2.  11 cm
    15 cm

3.  4 ft
    6 ft

4. **Sports**  Badminton is played on a rectangular court measuring 44 ft by 20 ft. What is the area of a badminton court?

5. Keith Witt is the world's fastest window cleaner! In 10 seconds, he cleaned 3 rectangular windows each measuring about 47 in. by 42 in. Estimate the total area he washed.

6. **Mental Math**  The area of a rectangle is 24 square inches. Its length is 8 inches. What is its width?

7. **What If**  The sides of a 5 in.-by-11 in. rectangle are doubled. Is the area of the new figure twice as large? Explain.

8. **Journal**  Explain how to find the area of this page.

**FOCUS ON SKILLS**                          Name_____

**Perimeter and Area**

The following shapes are drawn on centimeter dot paper.  Find the perimeter and area of each shape.

1.                          2.                          3.

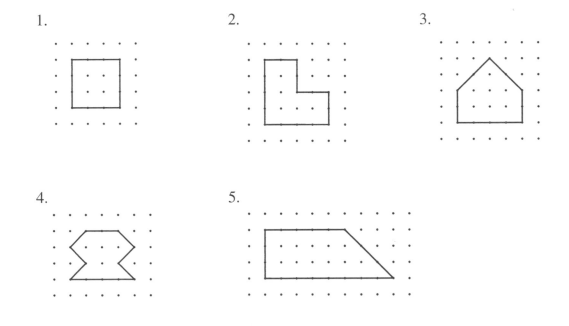

4.                          5.

6.  Find the perimeter and area of the triangle.

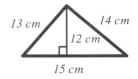

7.  Find the perimeter and area of the parallelogram.

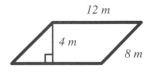

8.  Find the perimeter and area of a rectangle with a length of 6 inches and a width of 2 inches.

9.  Find the perimeter and area of a right triangle with sides 9 in, 12 in, and 15 in.

10. Find the area of a trapezoid with bases 11 feet and 15 feet and a height of 5 feet.

**FOCUS ON CONCEPTS**                          **Name**_____
**Perimeter and Area**

1. Using centimeter dot paper, draw all possible rectangles with a perimeter of 12 cm and whole number values for the length and width.

2. Give the dimensions of each rectangle and find the area.

3. Do any of the rectangles have an area equal to the perimeter? If so, describe the rectangle.

4. Which rectangle gives the largest area?

5. Repeat questions 1 – 4 for rectangles with perimeters of 16 cm and 24 cm.

6. A gardener has 20 feet of fence. He plans to enclose a rectangular garden with an area of 24 square feet. Does the gardener have enough fence to enclose the area? If so, what are the dimensions of his garden? If not, how much fence does he need to enclose the garden?

7. Mr. Strozzo wants to build a pen for his dog. The pen will have a concrete floor. He has enough concrete mix to cover a rectangular area of 36 square feet. What are the dimensions of the pen with the smallest perimeter that Mr. Strozzo can build for his dog?

8. Given a perimeter $P$, describe (give the dimensions of) the rectangle with the greatest area.

9. Extend your thinking beyond rectangles. Cut a piece of string 12 inches long. Tape the ends of the string together to form a loop. Stretch your string around pegs on a peg board to form various shapes. Find or estimate the area of each shape.

# VOLUME AND SURFACE AREA

Solid shapes are discussed throughout the elementary and middle school curriculum. The volume of such shapes is introduced in third grade, and surface area is taught in fifth grade. As students progress through the later elementary grades and middle school grades, these topics are expanded and reinforced. Eventually, students use formulas for finding volume and surface area of three-dimensional objects. As a teacher, you should be able to develop and explain the formulas, as well as use the formulas.

The sample pages for volume are from a Grade 4 book. Students are asked to determine the volume of a solid by dividing it into cubes one unit on each side and by using a formula. These activities help students understand the concepts of volume being measured in "cubic units."

The sample pages for surface area are from a Grade 6 book. Students are asked to draw "nets" for various three-dimensional shapes on centimeter grid paper and to find the total surface area of each shape. Also, they are asked to find surface area using formulas and to explain why surface area is measured in "square units."

The *Focus on Skills* problems ask you to find the volume and surface area of various prisms, pyramids, cylinders, and cones. You may accomplish this by using a particular formula or applying a certain strategy.

The *Focus on Concepts* problems encourage you to use your problem solving abilities and make connections, as well as distinctions, between volume and surface area. Given a specific volume, you will determine possible dimensions for a rectangular prism (box). Also, given a specific amount of material (surface area), you will determine what size boxes could be made. Finally, you will explore the same problem with a "closed" box.

**Chapter 8**
**Lesson**
**13**

# Exploring Volume

**Problem Solving Connection**
Use Objects/
Act It Out

**Materials**
■ color cubes
■ calculator

**Vocabulary**
**volume**
number of cubic units needed to fill a solid

**Units for Measuring Volume**
cubic unit
cubic centimeter
cubic inch

**Remember**
A rectangular prism is a solid with a right angle at each vertex.

**Explore** • • • • • • • • • • • •

The **volume** of a solid is the number of **cubic units** it contains. This is one cubic unit.

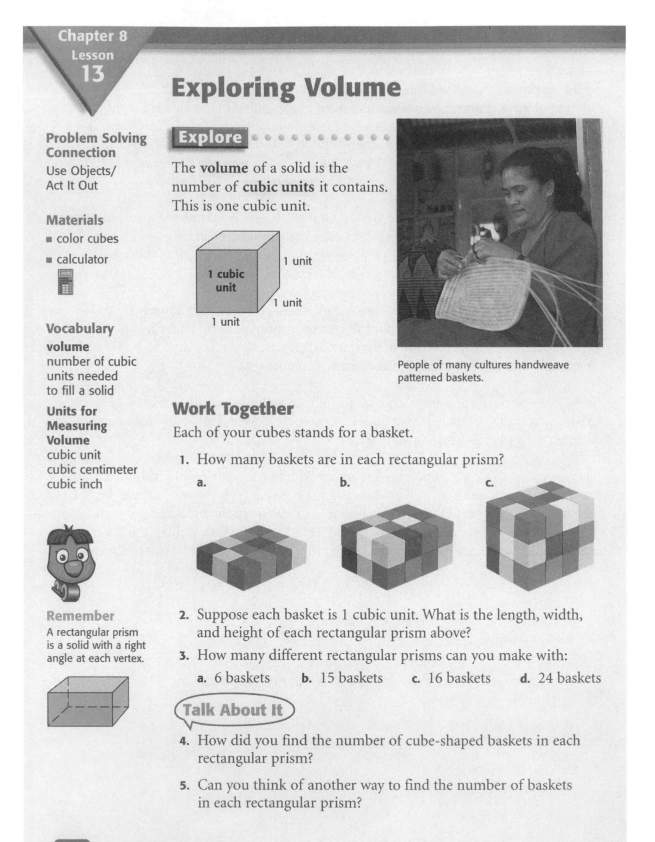

1 cubic unit
1 unit
1 unit
1 unit

People of many cultures handweave patterned baskets.

## Work Together

Each of your cubes stands for a basket.

**1.** How many baskets are in each rectangular prism?

a.                  b.                  c.

**2.** Suppose each basket is 1 cubic unit. What is the length, width, and height of each rectangular prism above?

**3.** How many different rectangular prisms can you make with:

   **a.** 6 baskets    **b.** 15 baskets    **c.** 16 baskets    **d.** 24 baskets

**Talk About It**

**4.** How did you find the number of cube-shaped baskets in each rectangular prism?

**5.** Can you think of another way to find the number of baskets in each rectangular prism?

## Connect

You use cubic units to measure the volume of a solid. You can count the number of **cubic centimeters** or **cubic inches** to find the volume.

The volume is 5 cubic centimeters.          The volume is 2 cubic inches.

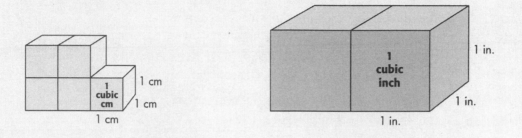

You can also find the volume of a rectangular prism by using a formula.

**Volume = length × width × height**

Volume = 3 units × 2 units × 2 units

Volume = 12 cubic units

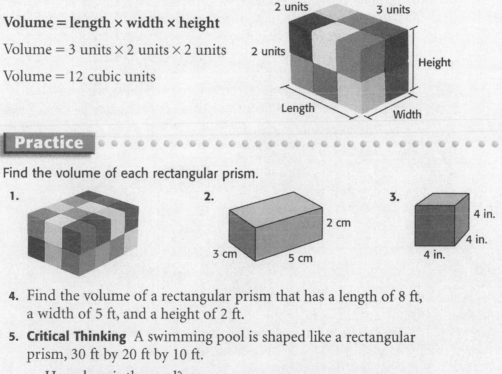

## Practice

Find the volume of each rectangular prism.

1. 

2. 

3. 

4. Find the volume of a rectangular prism that has a length of 8 ft, a width of 5 ft, and a height of 2 ft.

5. **Critical Thinking** A swimming pool is shaped like a rectangular prism, 30 ft by 20 ft by 10 ft.

   a. How deep is the pool?

   b. What is the area of the pool cover?

   c. How much dirt was removed from the hole to make the pool?

6. **Journal** Explain how to find the volume of a box that is 3 in. by 2 in. by 4 in.

# Exploring Surface Area

**Problem Solving Connection**
■ Draw a Picture/ Use Objects

**Materials**
■ Cuisenaire rods
■ centimeter graph paper

**Vocabulary**
surface area (SA)

## Explore ● ● ● ● ● ● ● ● ● ● ● ● ● ● ● ● ● ● ● ● ● ● ●

What does a box of dog biscuits have in common with its net? They have the same surface area!

The **surface area (SA)** of a polyhedron is the sum of the areas of all of its faces.

To find the surface area of a polyhedron, unfold it into a net of polygons. Then add their areas.

## Work Together

Use four Cuisenaire rods of the same color.

1. Create a rectangular prism by putting the rods together as shown.

   a. On centimeter graph paper, draw a net for this prism.

   b. Tell how many squares it will take to cover this prism.

2. Reassemble the same four rods as shown.

   a. On centimeter graph paper, draw a net for this prism.

   b. Tell how many squares it will take to cover this prism.

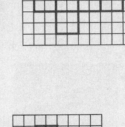

### Talk About It

3. Why must you be able to find the areas of polygons in order to find the surface areas of polyhedrons?

4. Which of the two prisms has the greater surface area? Tell how you know.

5. Why is surface area measured in square units?

6. Could you make more than one net for each prism? Would a different net give a different surface area? Explain.

## Connect and Learn

The surface area of a polyhedron is the sum of the areas of each face. You must first recognize the number of faces and their shapes.

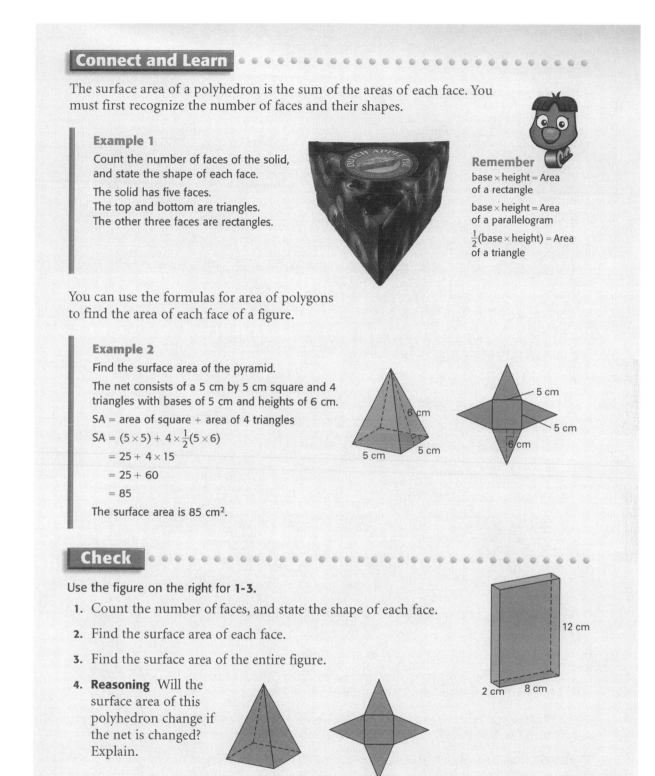

### Example 1

Count the number of faces of the solid, and state the shape of each face.

The solid has five faces.
The top and bottom are triangles.
The other three faces are rectangles.

**Remember**

base × height = Area of a rectangle

base × height = Area of a parallelogram

$\frac{1}{2}$(base × height) = Area of a triangle

You can use the formulas for area of polygons to find the area of each face of a figure.

### Example 2

Find the surface area of the pyramid.

The net consists of a 5 cm by 5 cm square and 4 triangles with bases of 5 cm and heights of 6 cm.

SA = area of square + area of 4 triangles

$SA = (5 \times 5) + 4 \times \frac{1}{2}(5 \times 6)$

$\quad = 25 + 4 \times 15$

$\quad = 25 + 60$

$\quad = 85$

The surface area is 85 cm².

## Check

Use the figure on the right for **1–3**.

1. Count the number of faces, and state the shape of each face.

2. Find the surface area of each face.

3. Find the surface area of the entire figure.

4. **Reasoning** Will the surface area of this polyhedron change if the net is changed? Explain.

## Practice •••••••••••••••••••••••••••••••

### Skills and Reasoning

Find each area.

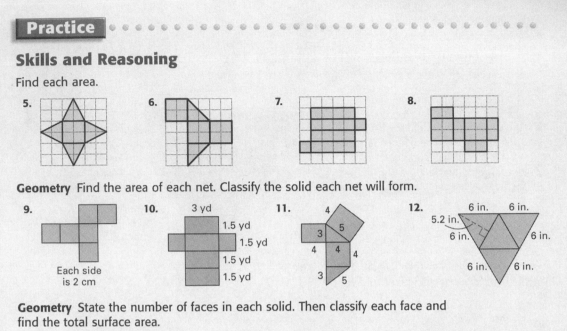

5.    6.    7.    8.

**Geometry** Find the area of each net. Classify the solid each net will form.

9.

Each side
is 2 cm

10.    3 yd

1.5 yd
1.5 yd
1.5 yd
1.5 yd

11.    4
3    5
4    4    4
3    5

12.    6 in.    6 in.
5.2 in.
6 in.    6 in.
6 in.    6 in.

**Geometry** State the number of faces in each solid. Then classify each face and find the total surface area.

13.    10 ft    7 ft
5 ft

14.    5 in.
4 in.    8 in.

15.    6 in.    6.4 in.
4 in.    5 in.

16.    Which net can form a cube?

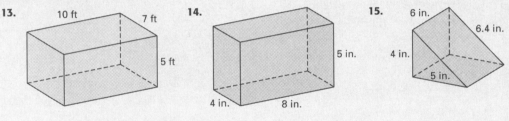

Ⓐ    Ⓑ    Ⓒ    Ⓓ

17.    If each side in a net for a cube is 3 in., what is the total surface area of the cube?

18.    **Calculator** Use a calculator to find the surface area of a square pyramid whose net consists of a 5 cm by 5 cm square and 4 triangles with bases of 4.6 cm and heights of 7.8 cm.

19.    **Reasoning** Will the surface area of a cube increase by 2 square units if the area of each face increases by 2 square units? Explain.

20.    **What If** Suppose the surface area of a cube is 6 square units. If you put two of these cubes end to end, would the surface area of the new polyhedron be 12 square units? Explain.

**Math Tip**

The M+ button adds numbers to the calculator's memory. Press
5 ✕ 5 = M+ to store the area of the square. When you know the area of the four triangles, press
+ MR = to add this to the number in the memory.

## Problem Solving and Applications

21. **Measurement** The United States Post Office only delivers mail that is at least 0.007 in. thick. If a piece is between 0.007 and 0.25 in. thick, it must also be at least 3.5 in. long and 5 in. wide. What is the surface area of the thinnest piece of mail that the U.S. Post Office will deliver?

22. Marie is making spaghetti for dinner. The box of spaghetti measures $1\frac{1}{2}$ in. wide, 4 in. long and 12 in. high. What is the total surface area of the package?

23. If wrapping paper costs $0.29 a square foot, how much would it cost to cover the box shown?

1 ft
0.3 ft    1 ft

24. **Critical Thinking** Regan has enough foil to cover half of the larger box shown. Since the dimensions of the smaller box are half those of the larger box, she thinks she can completely cover the smaller box instead. Do you agree with Regan? Explain.

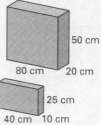

50 cm
80 cm    20 cm

25 cm
40 cm    10 cm

25. **Patterns** A manufacturing company wants you to design a box with a surface area 4 times that of its standard 2 in. by 2 in. by 2 in. box. Find as many whole-number solutions as you can.

26. **Journal** List some differences between the perimeter of a rectangle, the area of a rectangle, and the surface area of a rectangular prism. Can any of the quantities be negative? Explain.

## Mixed Review and Test Prep

For each ratio, give two equivalent ratios.

27. $\frac{3}{4}$    28. $6:24$    29. $\frac{4}{10}$    30. 7 to 9    31. $\frac{11}{12}$    32. $3:8$

33. $\frac{1}{9}$    34. $2:3$    35. $\frac{6}{10}$    36. $16:3$    37. $\frac{10}{15}$    38. 4 to 7

Order from least to greatest.

39. $\frac{1}{2}$, 0.23, 1.23    40. $6.7, \frac{1}{6}, \frac{5}{6},$    41. $8.2, 8\frac{1}{4}, 8.75$    42. $\frac{1}{3}, \frac{3}{3}, 3.3$

43. Which shape cannot be used to make a tessellation?

Ⓐ Equilateral triangle    Ⓑ Parallelogram

Ⓒ L-shaped figure    Ⓓ D-shaped figure

**FOCUS ON SKILLS**                              Name_____
**Volume and Surface Area**

Find the volume and surface area of each solid.

1.

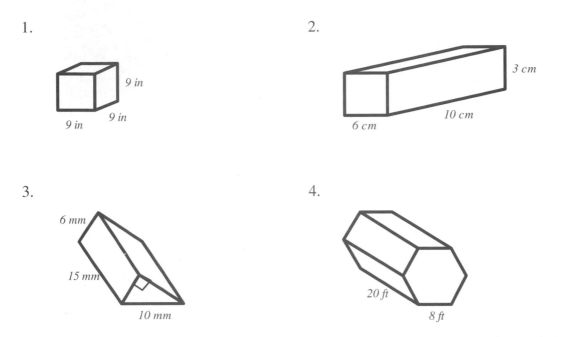

9 in
9 in
9 in

2.

3 cm
10 cm
6 cm

3.

6 mm
15 mm
10 mm

4.

20 ft
8 ft

(assume the hexagon is regular)

5.  Find the volume and surface area of a cube with a side of 15 feet.

6.  Find the volume and surface area of a right cylinder with height 12 inches and radius of the base 7 inches.

7.  Find the volume of a square pyramid with a base 6 cm on each side and a height of 18 cm.

8.  Find the volume.

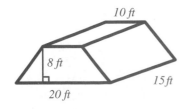

10 ft
8 ft
15 ft
20 ft

9.  Find the volume of a waffle cone with a height of 8 inches and a 4-inch diameter open end.

**FOCUS ON CONCEPTS**                          Name_____
**Volume and Surface Area**

1. Consider a box shaped as a rectangular prism. The volume of the box is 24 cubic feet.

   What are the possible dimensions (length × width × height) of the box if the dimensions are whole numbers?

   What is the surface area of each "possible" box (assume the box has no top)?

   What are the dimensions of the box with the smallest surface area?

   You may find it useful to organize your work in a table such as the one below.

   | Box # | Length | Width | Height | Surface Area |
   |-------|--------|-------|--------|--------------|
   |       |        |       |        |              |

2. You plan to construct an open cardboard box. You have 30 square feet of cardboard.

   What are the possible dimensions (length × width × height) of the box if the dimensions are whole numbers?

   What is the volume of each "possible" box?

   What are the dimensions of the box with the greatest volume?

   You may find it useful to organize your work in a table such as the one below.

   | Box # | Length | Width | Height | Volume |
   |-------|--------|-------|--------|--------|
   |       |        |       |        |        |

3. Suppose you want to construct a rectangular prism (i.e. a box enclosed on all sides) with the 30 square feet of cardboard. What are the possible dimensions (length × width × height) of the prism if the dimensions are whole numbers?